The Love of Nature and the End of the World

The Love of Nature and the End of the World

The Unspoken Dimensions of Environmental Concern

Shierry Weber Nicholsen

The MIT Press
Cambridge, Massachusetts
London, England

This book was set in Berkeley Old Style Book by Achorn Graphic Services, Inc. on the Miles 33 system.

Printed and bound in the United States of America.

Library of Congress Cataloging-in-Publication Data

Nicholsen, Shierry Weber.
    The love of nature and the end of the world : the unspoken dimensions of environmental concern / Shierry Weber Nicholsen.
      p. cm.
    Includes bibliographical references and index.
    ISBN 0-262-14076-4 (hc. : alk. paper)
    1. Human ecology—Philosophy. 2. Environmental degradation—Psychological aspects. I. Title.

GF21 .N53 2001
179′.1—dc21

2001044329

Arden H. Nicholsen
in memoriam

Contents

Acknowledgments

The stuff of this book is derived in large part from conversations with many people in many contexts over the past years—friends, colleagues, teachers, and students; people I have come to know in Seattle's psychoanalytic community, in Antioch University's Environment & Community program, in the arts, and elsewhere. There are many people to whom I am grateful for very specific contributions, of widely varying kinds, to the book. Among them are Barbara Barry, Kim Buehlman, Guy Burneko, Judy Cadman, Austin Case, Karen Coffey, Brian and Pam Costner, Susan Czopek, Francesca de Gasparis, Gene Dilworth, Jeffrey Eaton, Joan Fabian, Marilan Firestone, Janet Gerard, Raelene Gold, Terry Hanson, Caron Harrang, E. Lynn Hassan, Kim Hayes, Sally Hufbauer, Jeffrey Kahane, Katherine Knowlton, Angela Leja, Gaeage Moetse Maher, Peter Martynowych, Rachel Matthews, Steve Maurer, Betsy McConnell, Joan Mellon, Marllan Meyer, Randy Morris, Bev Osband, Phillip Penna, Janet Pfunder, Janet Sailer, Trent Schroyer, Jeremy Shapiro, Chris Slesar, Gene Sterling, Mark Windschitl, and Sara Winter.

I have been deeply touched by the thoughtfulness and eloquence of the students in my environmental philosophy courses. Many will find thoughts and feelings they expressed reflected here. I am grateful to them all.

My thanks to all the people who read earlier versions of the manuscript and offered their comments, which were helpful in ways they might never have anticipated. I am particularly grateful for the exceptionally close readings my friend Sally Hufbauer and an unnamed reviewer gave the manuscript and the detailed suggestions they offered.

I owe a special debt to my friend and colleague Stephen G. Shehorn, who not only graciously allowed me to accompany him on a hunt but also imparted some of the elegance of his own style to portions of the manuscript he edited.

This book could not have been written without the support and illumination afforded by my personal psychoanalysis—one of whose aims is to allow the unspoken to become speakable. For this I am deeply grateful.

The opportunity to present and discuss portions of this material in various professional settings has been extremely helpful, and I would like to express my gratitude to the groups and organizations that provided those occasions, among them Antioch University Seattle, the Northwest Alliance for Psychoanalytic Study, the Pacific Northwest Psychoanalytic Society, the Seattle Counselors Association, The Colorado College, and the Center for Western European Studies at the University of California at Berkeley.

And finally, I would like to express my gratitude to Larry Cohen and Clay Morgan of The MIT Press for the welcoming home they provided the idea that became this book.

Introduction

This book has its starting point in a persistent question: How can the public mind relegate matters of the environment, which is the ground of our whole lives, to the periphery of concern, as though they were the private interest of a group called "environmentalists"? At the same time, I have never met anyone who did not value and appreciate some part of the environment. How can we be so split in our thinking?

Only a few people have tried to answer this question. One is Harold Searles, a psychoanalyst. In 1972 Searles wrote an article, little known among either clinicians or environmentalists, called "Unconscious Processes in the Environmental Crisis."[1] "Even beyond the threat of nuclear warfare, I think, the ecological crisis is the greatest threat mankind collectively has ever faced," Searles wrote. He added: "My hypothesis is that man is hampered in his meeting of this environmental crisis by a severe and pervasive apathy which is based largely upon feelings and attitudes of which he is unconscious."[2] I consider this statement the basis of what I have tried to do in this book—to explore the psychological reasons for what appears as willful stupidity.

Psychoanalyst Wilfred Bion talks about the need for "binocular vision" in order to feel our participation in the irrational while being able to turn a curious and thoughtful eye on it. My intention here has been to bring together our sense of connection with the nonhuman environment—its beauty, its mystery, its provision of a sheltering home for us—with the psychological forces that allow the destruction to continue. I think each of us knows both, and it is important to acknowledge their conjunction in us.

With few exceptions, people writing about the natural environment and people concerned with the interior of the psyche have not drawn on each other's work. Among the exceptions are Searles, Robert Jay Lifton, and the Jungian James Hillman. I have drawn heavily on all of them. Paul Shepard is perhaps the only person to have focused fully and directly on the question of how human development relies on the natural environment, and how culture facilitates or disrupts this process. I have drawn heavily on all these writers, and their influence on my thinking will be evident to the reader.

The groundwork for my efforts has been laid in other areas as well. There is an abundance of insightful, elegant, and persuasive writing in the Western tradition that evokes and articulates the sense of connection with the natural environment, from Thoreau onward to Gary Snyder, and later David Abram and many others as well. It is now more a question of how to assimilate this vision than how to elaborate or add to it. Berkeley architect Christopher Alexander, continuing Kant's understanding of the aesthetic as a link between realms, has made a particular contribution to our understanding of the way beauty and aesthetic experience bridge and unify our human and natural worlds and our internal and external experience. Like the environment, he would argue, aesthetic experience is not a peripheral specialty but the ground of our human life. His work has been the impetus to include an exploration of the role of beauty here.

In my efforts to explore the conjunction of our attachment to and our destructiveness toward the natural world, I have chosen to proceed by evoking aspects of experience—primarily emotional and perceptual experience—in the reader's mind so that they can be reflected on in their complexity. I have taken certain themes in our relationship to the natural world and tried to illuminate some of their many facets and sound some of their various tonalities. The result might be described as a series of interconnected meditations, or perhaps, to use Theodor Adorno's term, a set of developing variations. In many cases I have taken as the starting points for these meditations the words of others who have thought deeply about these issues—those thinkers mentioned above, and many others, published and unpublished, as well. The polyphonic effect thus created also evokes, I hope, the complexity of the subject matter. In addition, it affords the reader a taste of the variety and richness of the work that bears on these questions, work that is for the most part familiar only to specialized audiences but that deserves to be more widely read.

In giving the book this particular form I have had to forgo many kinds of exploration that would have been useful to those concerned with my central question. I have focused on evoking experiences and elaborating central images and ideas, for instance, rather than discussing the empirical research that bears on them. This choice reflects my intention to open up issues rather than offer anything approaching full examination of them. The reader will need to look elsewhere for the details of research and the arguments around debated issues in these areas. Thus I have not tried to demonstrate that environmental degradation is in process but have simply taken it as a premise. More important, I have not discussed the research and the debates around key questions about indigenous cultures—questions, for instance, about variations in ecological awareness, hunting and gathering practices, or the role of ritual in specific cultures. Nor have I gone into important areas of research in environmental psychology, as for instance research around children's relationship to place and their sense of the natural world. Further, I have spoken of experience as though it were experience shared by a collective "we" that is the psyche of Western industrialized peoples. Empirically, this is problematic. Certainly there are wide variations in the conscious experience of Westerners, and certainly there are people in nonindustrialized countries who share some of these experiences. Nevertheless, it is my belief that despite individual variations the conjunction I am concerned with—the sense of connection with the natural environment and the psychological registering of environmental degradation—is fundamental to contemporary human experience. I am interested in evoking the reader's own version of that conjunction rather than proving that it exists or specifying the range of its variations.

The book's form will present a challenge to the reader. For one thing, few readers will find themselves on wholly familiar ground here. The book draws on work in the fields of environmental philosophy and ecopsychology, familiar to some, but also on the work of psychoanalytic thinkers like D. W. Winnicott, Donald Meltzer, and Wilfred Bion, whose thinking even professionals in the field often find difficult. Aesthetic theory and ideas from Buddhist and Sufi traditions are included as well. While I hope the book will inspire the reader to explore the work of the various writers I have drawn on, that is not necessary. Rather, I hope that the reader will take what is offered here as food for thought, an opportunity for things unthought and unspoken to be evoked by allowing the phrases to resonate in the mind and lead one where they will. At the same time, the book requires the reader to tolerate a certain degree of disturbance as he or she is unsettled by the juxtaposition of unfamiliar materials—materials that do, however, in the end present a surprisingly unified picture. And of course it is not only the unfamiliarity of some of the work I draw on that will be unsettling. The book attempts to evoke the emotional impact of environmental deterioration, and that impact is disturbing. A friend who read the manuscript commented that the book's premise was "we had better cry a lot," and he may have been right. I do think we have a lot of mourning to do. The importance of mourning notwithstanding, my primary intention in the book is to raise issues and evoke feelings in an attempt to elicit reflection on our terrible dilemma rather than to offer answers and solutions. That too is unsettling and requires tolerating uncertainty while the process of individual and collective reflection proceeds.

In keeping with the book's form, each chapter stands largely on its own, and within each chapter I move from one facet of an issue to another. Yet the reader will be aware of multiple links between the various sections and between the chapters. Sometimes those links are made explicitly and sometimes they are provided by allusions. Sometimes they are implicit in the overall sequence of the material. I begin the book with our sense of connection with the nonhuman, hoping to establish this in the reader's experience before moving to the questions of apathy and destructiveness. The following outline of the chapters may provide a helpful guide.

Chapter 1 deals with the issue of "the unspoken." It evokes the many and various reasons for silence and the process whereby we move from the unspoken to the spoken. It hopes to lead the reader to acknowledge that unspokenness is a vast and important territory and that it is particularly relevant to our relation to the natural environment.

Chapter 2 deals with "the love of nature" in its various forms and the question of how concern might arise. It is based on the assumption—in which I concur with Harold Searles—that the nonhuman environment is an important presence for each of us from the beginning of life. It hopes to evoke that recognition in the reader, and then to lead the reader into the sometimes-bewildering question of the nature of our relationship with the nonhuman.

Chapter 3 elaborates a possible relationship with the nonhuman expounded by David Abram and others that I call *perceptual reciprocity*. In the background is Paul Shepard's argument, convincing to me, that cultures in which such perception is possible allow for a greater degree of maturity. Thus I coin the term *perceptual maturity*. Cézanne's paintings on the one hand, and the hunt on the other, are used to explore that kind of perception.

Chapter 4 deals with beauty, which is found both in the human realm and in the natural world and as such serves as a way to think about the continuity between them. Here I bring in the work of Christopher Alexander, particularly through his less well known book on early Turkish carpets, *A Foreshadowing of 21st Century Art*. Ideas from the Buddhist and Sufi traditions appear here in connection with the question of beauty.

Chapter 5 moves to the other side of the relationship with nature—the "end of the world," that is, the psychological impact of environmental degradation and the destruction of the natural world. It uses work on the effects of trauma to elucidate the perhaps subtler effects of environmental degradation. It draws particularly on the extensive writings of Robert Jay Lifton and others on the psychology of events and situations like Hiroshima, nuclearism, and the Nazi Holocaust.

Chapter 6 reflects on the issue of the future as a way of asking about the implications for action of all that has been said here. It is particularly concerned with our capacity to think about time and the future, and about issues of leadership and group and communal life as we move into the future. Failure to deal with those questions would mean continuing as we are. Wilfred Bion, who in my opinion is the most original and incisive of psychoanalytic thinkers, emerges as the major figure of this chapter.

Bion hoped we would "dare to face the facts of the universe in which we live." The book as a whole is clearly more concerned with elucidating the forces that would *interfere* with our capacity to conceive a future than with offering suggestions as to what we should do next. Yet the book does suggest certain relationships between thought and action, and by doing so offers an implicit perspective on the question of what is to be done about our situation. In the brief section titled "Concluding Thoughts," I have offered some reflections on the book's implications for that question.

Notes

1. In Harold Searles, *Countertransference and Related Subjects,* 228–242.

2. Searles, "Unconscious Processes in the Environmental Crisis," 228.

1   Many Silences

Our most passionate feelings—our most intimate loves, our most overwhelming fears, our most heartrending griefs, our blackest despair—are these really
spoken, even to those we love and trust? Are they really heard? Certainly not
in our abundant talk of environmental matters, argued so cogently and so fero-
ciously, documented so carefully, denounced so righteously, described so beautifully. Yes, they are difficult to put into words—more so when they concern
not the human only but the natural world as well. But we sense they are
widely shared. In the urgency of our situation, this speechlessness is mysterious. In hiding the depth of our concern from others, perhaps we also hide it
from ourselves. Would it make a difference if we were able to be more courageous in speaking it?

Why are we most reluctant to speak what is most important to us? Because
full speech exposes us fully. In speaking the fullness of our loves and our fears
we find ourselves suddenly outside the enclosure of privacy. This vulnerability
lies at the root of our speechlessness. Our loving binds us intimately with people and other life forms, and in opening to love we know that the objects of
our love are vulnerable as well. Thus we are vulnerable to partial or complete
loss. We can lose what we love. We can lose our own capacities to function.
We thus become vulnerable to our own desire to avoid pain and to the collective forces of denial and destruction.

We are silent with others so that our vulnerability, and the vulnerability of
our loved ones, will escape their notice. Our loves remain unspoken. Even alone
we are silent in an effort to ignore our vulnerability. Certain things thus become

unspeakable. But this unspeaking itself is experienced as a silencing, an intimidation. Intimidated into silence, we are deprived of the intimate voice that speaks fully of our precious loves and nightmarish fears. Mute, we pass our days in private retreat, renouncing the engagement that speech might allow us to have.

Love Protected by Silence

> *The wildness of the savage is but a faint symbol of the awful ferity with which good men and lovers meet.*
> —Henry David Thoreau, "Walking"

> *Mise mono ja nai.*   [These things we do not show to the people.]
> —Japanese phrase, on keeping rituals off-limits to tourists

Much of our intense feeling for the natural world is a form of love, based on intimate encounter. In some essential way, words fail when it comes to love. It is an emotional, sensual, aesthetic experience, best expressed, we imagine, by pointing to the beloved thing as though its mere lovely existence speaks for itself. "Look, look!" or "I love it here" may feel like the best words can do.[1]

But there are other reasons why we do not speak of our loves. What we love, we want to protect, as environmentalists are fond of noting.[2] Though threats to what we love often impel us to speak out, we also protect our loves by not speaking of them. To speak may be to invite harm. Why does a fly-fisherman refuse to divulge the location of his special trout stream? To protect it, in many senses.

Love demands privacy. It is guarded, fiercely, by a boundary of silence toward what is outside. Those feelings are the private domain of those who feel them for one another. When the private intimacy of the love relationship spills over the boundaries, there is a sense that it is dissipated, or sullied, contaminated by the less sensitive energies of the group outside. Love might then be talked about by nonlovers, and it disappears in such talk. The group and the couple, as Otto Kernberg says, are antagonists.[3] Love fears indifference and scorn. Who would want to expose their love, thing of beauty, to rejecting eyes? Perhaps we fear that the attitude in those eyes would remind us of our own capacity for indifference.

Love also fears envy. We know that love is precious, and we are lucky to have it. Do not tempt the gods, happy lovers! To speak of love is to expose it to the forces of envy, and envy often wants to denigrate and spoil. Perhaps we sense that some of the destruction of wild places is done with a feeling of vicious triumph over those who love them. Love guards against jealousy as well. Those we love might feel abandoned or betrayed by our love and loyalty to others. Could a beloved human tolerate knowing how much we love a particular place or creature?

The group too is jealous of lovers. Think of all the measures society takes to prevent the couple separating from the larger group: the wedding as a group function, for instance, with tin cans attached to the departing car of the newly-weds. Do not forget us, the rattle of the tin cans says; we have not forgotten you. Even when the love of nature is shared, as in groups of people going into the wild together, all individually loving the natural world, members of the group still feel the tension between being alone with their loves and maintaining camaraderie. Indeed, what we speak about to others may be not a question of expressing or shielding our love so much as a question of the group's implicit rules for demonstrating solidarity. The country person who hunts and fishes and the activist who defends the forest may allude to their very similar loves in wholly different ways.

The question of speaking or not speaking our loves brings our dependency and vulnerability to the fore. Love is fragile. In love we are dependent on what we love. It can be lost, or harmed, and it can also betray us. We love not only the natural world but also other human beings, and we are dependent on the human community as well as on our private loves. Human beings too can be lost and harmed, and human beings can betray and exclude us. We hope to belong more deeply, but in caring we risk betrayal and loss.

That Way Madness Lies: Hiding Our Fears and Anxieties

"Not to see it, *the terrible* it!"
—Pierre Bezuhov, in Tolstoy's *War and Peace*

Some [physiologists] adopted a routine precaution: at the outset of an experiment they would sever the vocal cords of the animal on the table, so that it could not bark or cry out during the operation.
—Neil Evernden, *The Natural Alien*

The physiologist who cut the vocal cords of the animal he was vivisecting, Evernden tells us, was both denying and affirming his humanity: "He was denying it in that he was able to cut the vocal cords and then pretend that the animal could feel no pain, that it was merely the machine Descartes had claimed it to be. But he was also affirming his humanity in that, had he not cut the cords, the desperate cries of the animal would have told him what he already knew, that it was a sentient, feeling being and not a machine at all."[4] To feel the desperate suffering of any creature is terrifying. It can be so terrifying that we want to shut it out of our awareness. We ourselves are prey, sometimes consciously, sometimes beneath the surface of our awareness, to a host of desperate anxieties—about the holocaust of nature, the collapse of the world, the failure of a future. These we leave almost wholly unspoken. "It's scary," someone will say, and then be silent. It is as though we are cutting our own vocal cords.

One reason we keep silent about these profound anxieties is that we are afraid of "losing it," going mad. Think of the silliness of soldiers on leave from the front. They simply cannot bear to contemplate their situation as a whole. That way madness lies. What madness is that? These are the dreadful anxieties first encountered in early infancy: the fear of annihilation, of utter loss of orientation, of abandonment by any kind of caretaker or benevolent authority who can take responsibility and protect us from destruction. Fears of wholesale environmental destruction are of precisely this kind—states of terror combined with utter helplessness and dependency. The more desperate we feel, the more we wish to ignore these fears, to keep moving straight ahead without looking to the side. Susan Griffin characterizes this state of mind in words that evoke the urgency of the denial that pervades it:

> Yet perhaps it is the very extremity of the danger, bordering as it does on the continuity of life itself, the desire for safety as an ultimate state that seals away all fear as if into a foreign country, the wish for a miraculous, mysterious security won not so much by practical effort, or seen through theoretical understanding, but by a determination to keep on in one direction despite every indication of trouble, hence vanquishing not only this danger but all catastrophe and every mortal mistake by a sheer act of will, a terrible fear of danger that causes this denial of danger.[5]

There are good reasons to fear loss of the capacity to function. In the grips of these anxieties, we are utterly dependent on other people, just as the soldier who goes berserk and wants to rush out into fire needs to be restrained by his comrades. This degree of blind dependency is terrifying to contemplate. Our fearful fantasies include our anxieties about whether we will be able to continue to function on behalf of others. In a state of utter helplessness we cannot function as adults and be of help. Parents are desperate about their desperation because they know that their children need them to help with the children's own desperation.

This urge to "keep it together" can also mean "keep up a good front." It is reinforced by group pressure. Panic is contagious, and the group is afraid of being infected by it. The group wants to continue to function, and fears the consequence of having its anxieties evoked. A good front, however, alienates people from important aspects of themselves and one another; it is a shared falseness within which it is hard to move toward facing reality.

The consequences of feeling and speaking one's deepest fears do indeed seem terrifying. Yet I believe it is this shared fear of madness that deprives people of the opportunity to think collectively about their fears by putting them into words.

Mute Shame and the Silence of Catastrophe

To live without words is a terrible thing.
—Joan Mellon

Was it not noticeable at the end of the war that men returned from the battlefield grown silent—not richer, but poorer in communicable experience?
—Walter Benjamin, "The Storyteller"

The group has strong weapons to keep its members from leaving it for the private domain of love—shaming and scorning, for instance. It is taboo to express one's feelings for the natural world too strongly. We might be called anti-human, immature, unhealthy, obsessed, not to mention greenie, nature-lover, tree-hugger, weirdo, kook. . . .

A friend of mine grew up on a farm. His family was not close and offered the boy little emotional support. He was an anxious child and spent a good deal of time alone in the fields and the woods. The trees were reassuring presences to him. They endured, season after season, and each one seemed an individual. Animals were more frightening: they made loud noises and could move fast and get out of control. They gave birth and were killed and eaten. But all those experiences of the natural world remained wordless. Who would he have talked to about them? There was no hearing for a child's emotional experiences in that family. Even if he had had words to express them, he would have felt silly doing so. They were part of a private world. Words were for the public world of school, but there was no talk of feelings or anxieties there. To try to bridge the divide between public and private, he felt, would be to risk unbearable shame and scorn.

Experience that is unbearably painful is impossibly difficult to communicate, and one falls mute. As Susan Griffin remarks, "a certain kind of silence is a common effect of catastrophe."[6] The very fact of not being heard gives rise to a shame that is further silencing. The more violently painful the experience, the more abusive and traumatic the lack of reception, the greater the muteness and the shame. This is why war leads to so much muteness. War uses up words, as Henry James said. What happens then to the experience? It is as though the not-hearing is taken back into the self and becomes a barrier of silencing turned inward, shutting away and even erasing the experience itself. Men return silent from the battlefield, poorer in themselves.

There is a relief that comes with speaking, and with it a potential for growth, understanding, and effectiveness. Something changes when things are put into words. Suffering becomes minimally more bearable, at least open to reflection. When there is no one to hear our feelings and no one to speak the words for them—words that help our minds formulate their own words— our experiences do not become experiences. It is as though they are not contained within the part of ourselves that can think and speak and make decisions. Just as certain sensations are intensely disturbing, though we cannot easily explain why, there are internal, intangible experiences that we cannot take physical hold of, all the more terrifying in that they seem to *be* us. To try to make contact with these not-yet-experienced experiences is immensely anxiety provoking. We fear that we will once again experience not being understood, not being helped to make meaning and words. We fear that the grief will be overwhelming, the anxiety will fragment us, the images and fantasies will be overpowering. "I could bear it no longer, and lost consciousness," wrote a soldier of his agony in the war.[7]

The muteness into which we are plunged by what cannot be communicated is also the silence of others' experience withheld. This is the corrosive effect of shameful secrets. They deprive us of the food of truth. "The soul has a natural movement toward knowledge, so that not to know can be to despair," writes Susan Griffin. She adds: "A certain kind of silence, that which comes from holding back the truth, is abusive in itself to a child."[8]

Abuse is shameful and silencing not only because of the victim's pain but because it destroys the sense of relatedness among those who do not acknowledge it. A friend told me this story. He said he had never told anyone before. As a boy, he used to visit his "cowboy" grandfather during the summer, the one who carried a Magnum in case he encountered a rattlesnake. He feared his grandfather and dreaded those summer visits. Going fishing for catfish in the cowpond during those summers was part of the ritual of coming into manhood his grandfather's way. One day his grandfather landed a big catfish and asked the boy to grab it. The boy let it slip by mistake, and it escaped into the pond. He was ashamed and cried; he was failing as a man. He tried to make up for it by catching a catfish himself, and he did catch one. Then it had to be cleaned. His grandfather's way of cleaning a catfish was to nail the snout of the living fish to a board and then pull the fish's skin off with a pair of needle-nosed pliers. The fish the boy caught must have revived a little during the process, for it screamed—the horrifying, chattering, unforgettable scream of a creature in agony. The boy could never forget that scream, or ever make up for what he had done by catching that fish, or ever speak of the experience to the one he had shared it with, his grandfather. The scream of agony was matched by the silence and shame in which the experience was buried.

Sometimes the bond with the natural world is forged through suffering. The scream of the catfish is received by the boy, who knows suffering. But how loud and how excruciating is this suffering that cannot be acknowledged and talked about! And how strong a role fear plays in this unspokenness. The boy is afraid of his grandfather, who has shown how he can deal out pain to living creatures, and he is afraid to acknowledge this experience of shared suffering, of which he and his grandfather are the witnesses. And the pain in the grandfather, which led him to be so hard? Unspoken, the food of truth denied, the child condemned to silent shame.

The Voice That Comes to Speech

> *I fell ill, broken in body and spirit. The main symptoms of my illness were fits which began by my hearing the voices of my fellow soldiers*

becoming louder and louder until I could bear it no longer and lost con-
sciousness.
—Alfred Wolfsohn, *The Problem of Limitations*

In her book *A Chorus of Stones: The Private Life of War,* Susan Griffin writes not only of the effects of silence and secrets and the shame of abuse, but also of the soul's hunger for revelation and truth, for voice to be given to what has been kept hidden and denied. "Children who have been abused will turn inward, but there is something that will wake them," she writes. And "that is if the abuse to which they have been submitted is named and admitted to be true."[9]

Griffin writes of Alfred Wolfsohn, a Jew who served as a stretcher bearer in the German army in World War I. He was wounded and buried under rubble among the dead. Afterward he suffered from what was then called shell-shock. Those experiences became remarkable in Wolfsohn's experiments with the human voice after the war. As a most unusual voice teacher later in his life, Wolfsohn opened up the range of the voice, extending it by octaves, letting it speak for the whole of the body's experiencing and suffering. This was the eventual fruit of his catastrophic experience in the war. The voice that he found—"unchained" was his word—was to his mind the voice of the soul, which spoke of life, and the voice of God.

Wolfsohn's way to his discovery was through the voices of the dying, voices of agony and unbearable suffering. "In the cries of the dying, he had heard a range of the human voice beyond all conventional expectation," writes Griffin.[10] Those cries came to him in an experience of his own agony, mixed with guilt and shame. For Wolfsohn turned away from the dying's pleas for help, and in his mind his refusal to hear those voices was the cause of his illness. He had turned away from the voice of God as he crawled through the battlefield:

I had turned a deaf ear to this voice when I crawled hour after hour, inch by inch, in the trenches, haunted by the horrors of hell, cursing and denying God. Somewhere, someone cried out: "Comrade, comrade!" a fellow creature, writhing in agony like me. I was terror-stricken. "You must crawl to him," my inner voice shouted—"No, you cannot, you must save yourself. Who helped you? Your comrades also passed you by." And I crawled on, was buried under rubble and awoke amongst corpses. It was then, it seems, that I lost my soul. However I have not forgotten the soldier's voice which would not leave me, penetrating me deeper and deeper, poisoning my whole being.[11]

The voice that stayed in Wolfsohn's mind, poisoning it, as he says, eventually brought him back to the true voice, the voice of life. In an experience so potent as to be at first unbearable, he received the voice of agony into himself, and it yielded an immense truth. Wolfsohn's later work with what he called "singing" showed that agony can be given voice. It is as though the boy himself could have screamed the scream of the catfish and put it into words in such a way that his grandfather could have heard and acknowledged the suffering.

A woman I know grew up in the country, on a farm, but the discord in her family was so great that she had no attention for the countryside. She merely wanted to survive her childhood and escape to the city. As an adult, she became quite sociable and had many friends. One day she saw a movie about a heroic lawyer and a polluted river running through a poor community. She found herself weeping uncontrollably for days afterward. She was weeping, she told me, not over the heroism or the suffering of the community, but over the tragedy of the polluted river—"the poisoned waters," she called it. A few months later, staying at a friend's house in the woods, she found herself sitting silently and alone by the pond every morning, marveling at what she had never noticed before.

This woman had had abundant words for other people. What she had not had was the capacity to be alone with herself in solitude and silence. The silence she came to was not so much speechlessness, the absence of words, but an ability to let people fade into the background while she entered into a different kind of relationship with the pond. But to come to the silence, she needed first to come to speech. She had had no interest in the natural world growing up because (or so I imagine) her head was full of things she needed and wanted to say to the adults she was living with, things she could not say because of all the discord in that family. Something had to happen to allow her to acknowledge the poisoning in her family and give voice to her grief about it. That something, I believe, was the film. When the film, and the hero in it, acknowledged the suffering of the river and the people who lived by it, she became able to acknowledge and cry out her own agony. Afterward she was free to find a different kind of silence.

The Awe Before There Are Words

> *The roaring of lions, the howling of wolves, the raging of the stormy sea, and*
> *the destructive sword are portions of eternity, too great for the eye of man.*
> —William Blake, *The Marriage of Heaven and Hell*

To move from speechlessness to speech requires a person—perhaps a wiser part of ourselves—who can hear and receive our experience. As we are heard, we become able to hear our experience ourselves. In the beginning, however, is speechlessness, unformed experience no doubt both beautiful and terrifying. Silence sometimes means that there are no words yet.

Awe touches us even more deeply than a felt love, yet it is deep in darkness. It is not simply unspoken; it is speechless. A friend tells me that she cannot describe her feeling for the natural world as love. It is not love but awe, she says. She is simply struck speechless at the sight of a heron lifting its wing. Awestruck, she is incapable of saying more.

In part, awe does not have words because it is utterly private, not "for show." But awe is more than private. It is an involuntary speechlessness. That we seldom find the sense of awe in our talk about the environment may be due in part to a our diminished capacity for awe, but it is also due to the inherent speechlessness that awe brings us to. We cannot even put words to it ourselves. It is not surprising that we do not speak of it to others.

Awe is the sense of an encounter with some presence larger than ourselves, mysterious, frightening and wonderful, numinous, sacred. It is the sense of something that we are not capable of containing within our capacity for thought and speech. In awe, one's self is felt only as something small and incapable, speechless, perhaps graced by the experience but unequal to it, humble.

Awe makes us feel amazed, astounded, struck dumb. Joseph Campbell's term *aesthetic arrest,* which denotes something similar, conveys this sense. We are stopped in our tracks. The words *amazed* and *astounded* both suggest a blow to one's normal mental functioning, as when one is literally stunned or struck or loses one's normal orientation (as in a maze). Donald Meltzer tells the story of a little boy whose therapist, in a gesture out of the ordinary, wiped his face. The boy sat there "amazed." How are we to understand this? Meltzer quotes from the Jewish *Book of Law:* "Stand close to the dying, because when the soul sees the abyss it is amazed." For the soul of the one dying, death seems an "unbearably new" experience. When a particular emotion has never been felt before, it will not immediately yield its meaning, says Meltzer, and the psyche responds with amazement.[12]

The notion of an experience that does not immediately yield its meaning is the key to the speechlessness of awe in the face of the natural world. While awe stops us in our tracks, this is not the end of our experiencing but rather a beginning. Somehow we intuitively sense that the experience is extraordinarily rich and will require vast transformations of our mental structures as we assimilate it. Our intuition of the long, unknown process ahead stops us cold.

A friend told me that when she first saw Cézanne's painting "The Bather" in the Museum of Modern Art in New York she experienced the figure in the painting as emerging from the canvas and coming toward her, then moving back, then emerging again. She was awestruck by Cézanne's capacity to shape space in this way. "It was almost like a bolt of lightning went into me," she said in recounting the experience. "Some mystery, something entered into me so strongly that it shook me to the bone." Overwhelmed, she burst into tears and ran out of the museum, feeling she could never do anything comparable. But in the decades since then, that experience of awe has entered into her own unfolding as a painter. "The Man Emerging from Space" is her own name for Cézanne's painting. In her story we see the painter emerging from the experience of awe.

What allows something in us to develop the potential inherent in the experience of awe is the same process that allows the infant to develop—things are put into words, or otherwise symbolized, by a mind spacious enough to hold or contain the experience. In Wilfred Bion's model of mental development, the mother, identifying with her infant's experience in reverie, "holds" it for the infant and gives it meaning in her own mind before communicating it back to the infant. Through this process the infant becomes able to symbolize experience and eventually put it into words. Emotional experience thus becomes usable for growth and development.[13]

As the evolutionary energies of the unbearably new experience do their work, we become capable of putting the experience into words. "I have never told anyone before, but I think you will understand," Kathleen Raine's mother said to her, as a preface to telling her an experience of awe.[14] She recognized the opportunity—a person who would "understand"—and she made the decision to put the unspoken into words. As we are heard, we become capable both of hearing ourselves, of making the effort to put things into words, and of recognizing a person who can hear. Courage and receptivity develop together.

It may take years for awe to bear its fruit, but the self holds something in readiness for a future time. The boy who heard the catfish scream must have needed to scream himself. But who would have listened? He listened to the catfish, who screamed instead. Yet his self held that experience until it could be told.

The process initiated by an encounter with the numinous does not occur without suffering. In some ways, awe and catastrophe are not so different. The experience of awe is so acute because we are at the very edge of awareness, at the edge of our capacity to tolerate the encounter with something so much greater than our present selves, at the edge of our capacity to tolerate our own humility and weakness. The unbearably new with its potential remains unbearable until it has been assimilated. How long could my friend bear her encounter with Cézanne's "Bather"? She burst into tears and ran out of the museum.

Jung claims that the incarnation of the greater Self in the individual self occurs through suffering.[15] The Bantu peoples of Africa, Laurens van der Post tells us, say that unless you find a way to take your tears and add them to the string of tears around your heart, your suffering will be in vain.[16] Finding the meaning of suffering is the process of finding the words for the experience one had, inchoately, in awe—but that process is none other than the process by which one develops the capacity to put experience into words.

Losing the capacity for awe means speechlessness of a different kind—the muteness that signals an incapacity to suffer. The self needs a measure of strength to bear awe and its attendant suffering, strength that comes from the experience of being heard and the struggle for words. Without that strength, it is not possible to make use of the experience of awe. The intense numinous energies encountered simply overwhelm and shatter the fragile self. Or the self defends itself by making sure never to come into contact with those energies. Suffering will then either be avoided altogether, or it will miscarry and turn to bitterness, cynicism, and despair.

The loss of the capacity for awe is not necessarily a permanent loss. As Susan Griffin insists, the desire for emotional truth remains, however hidden under silence and muteness. What is held unspoken contains the potential not only for great suffering, but also for numinousness, meaningfulness, and the evolution and development of human beings, individually and collectively.

Nature's Silence and Ours

> *A mysteriously living silence . . .*
> —Colette Richard

The natural world is more than and other than simply human. Our feelings of love and attachment for the natural world are for nonhuman creatures and for places. The anxiety we feel is not merely for the destruction of human lives but also for those other creatures and places, and for a world in which we would be at home. The awe we feel is evoked by a profound encounter with something other than ourselves.

Not only are these other beings and life forms not human, they are without human language. This means that our relation to the natural world is in some important way nonverbal and unspoken. We may speak to other human beings or to ourselves about our encounter with the natural world, but the encounter itself does not transpire in the medium of human language. Does this mean that to speak about that encounter is to objectify it rather than to express our experience directly? How indeed do I express and live my relatedness to the nonhuman? The absence of language puts the self into question. Who am I when I am with the nonhuman? Does speech belong to a different self than the one that relates to the natural world? Our wordless encounters with the natural world require and invite our capacity for silence and solitude. The self that goes alone to the woods where there is "nothing to do" is a self that endures even when not directly engaged in human pursuits. What is that self? Is there a sense in which it too is "nothing," and what then is this mysterious nothing?

Falling Silent

> *It is not birds that speak, but men learn silence.*
> —Kathleen Raine

To "learn silence," in Kathleen Raine's words, means that one must be able to become silent, to leave one's habitual talking. A friend told me a story of students who spent time in the Australian rainforest. Returning to town after a long period in the forest, her students said that in the forest they felt obligated to be silent, and on returning, they found the constant chattering of humans draining. What is it for, they asked themselves. Meditators returning from retreats report something similar: How tiring it is to talk!

Immersion in the great forest imposed a respectful silence on the students, which in turn allowed them to attend to the life processes of nature, the constant emerging of life in its own time, without artificial justifications. The function of much human talk seemed obvious by contrast—a blind way of proving to oneself that one is alive, or perhaps a way of putting barriers between oneself and another person, ensuring that nothing can emerge from the space that silence would make. Merleau-Ponty commented that looking at other paintings after Cézanne's is like hearing ordinary talk start up again after a funeral.[17] Entering that silence does indeed feel like a kind of death to the ordinary self.

The silence of nature is not a silence of individual creatures but the silence of the natural world as such. It is utterly compatible with sound. It is simply that the sounds of nature are contained within its silence. "Even the subdued roaring of the waves cannot destroy the proofed silence of the oceans," writes Omraam Mickhael Aivanhov.[18] This silence of nature is more than an absence of human language. It is an overarching sense of both containment and potential, of vitality ever emerging and not yet grasped. "Up in the heights the silence is white and luminous, a poetry of sunshine, glaciers, keen air and the scent of snow. . . . Underground it is a mineral stillness, immovable, a mysteriously living silence," writes mountaineer Colette Richard.[19]

Nature's silence is felt not as an absence, a lack of communication, but as a powerful summons on the part of the natural world, a demand for attention to be paid. Obeying this command means shifting to a different kind of attention. This shift is what we call "falling silent." We descend—really, we deepen—into a profound attention from which something can come to meet us. In this state, we can be touched by the intense aliveness and presence of the natural world and its creatures, among which are we ourselves. Laurens van der Post tells about a time when he and his companions entered Hokkaido in a storm and went into a restaurant for warmth. Suddenly, from the corner of the room behind him, he writes,

> a bird began to sing, and it sang with such beauty and such clarity and such authority that the whole room went silent. I have never heard a bird

song more beautiful. Both my friend and I were almost at once in tears.
. . . I do not know for how long the bird sang, but the silence was unbroken. Not a teaspoon made a glass or cup tinkle, not a whisper, not a
clatter of crockery or trays being laid out in the great kitchen came there
to disturb it. . . . It was a moment utterly timeless in a way that could
not be misunderstood, because it was free of all physical and material
barriers and impediments of personal pain and injury, as if it were fulfill-
ing directly the measure of the will of creation invested in that little body
of a small bird, un-wounding itself there and regaining its full sense of
being, with its heart in its throat.[20]

In the silence into which they fell, the men were able to experience the "full
sense of being," with its vulnerability (the bird had been blinded on the assumption that blindness would improve its singing), its beauty, and the authority inherent in life itself:

The immense power of the music had an almost paranormal quality of
command that was supreme, because it was not an expression of power
itself but came purely from what the music was within itself; an expression rather of ultimate harmony and beauty, asserting itself in its most vulnerable and defenseless form, relying for its own authority and impact solely
on its beauty and its necessities of order and measure and the lucidity of
its voice.[21]

And in a similar experience, in a restaurant in Kenya on a day full of dust,
van der Post tells us, the roar of a lion came suddenly from outside:

At once all conversation ceased and everyone listened with instinctive rev-
erence as if to the voice of a god. The lion was close, and the immediacy
of the sound came like lightning from its throat; the authority of the
voice proclaimed as if on behalf of life itself, through the absence of fear
and doubt in its utterance. Even when the lion's announcement ended,
we remained silent long enough to hear another lion answer at length
from far away. Only when that answer ended the primordial dialogue did
the men gasp, as if coming up for air out of an unfathomed deep themselves, and start to talk again.[22]

The commanding sounds of the bird and the lion plunge the listeners into an "unfathomed deep" of profound silence. There the voices of the nonhuman meet the receptivity of the listeners at a depth that does not have words. To those receptive ears the voices are those of a god, of life itself. The men are witness to a "primordial dialogue." This is an experience of awe, and there are no human words for it while it is happening. Indeed, when the people in the restaurant begin to talk, it is a sign that they have emerged from the depths and are no longer participating in the experience. If the words for this unfathomed experience are to be found at all, they will evolve over time, with the fathoming that is reflection.

The Strange Phosphorescence of Life: Solitude, Selfhood, and the Void

> *My mother when she was over eighty confided to me an experience she had had as a girl. "I have never told anyone before," she said, "but I think you will understand." It was simply that, one day, sitting among the heather near Kielder, "I saw that the moor was alive."*
> —Kathleen Raine, *The Land Unknown*

To enter the silence of nature is to abstain from the confirmation of human contact and speech. One sinks down into the abyss of the nonhuman. What one finds in nature's silence, however, is a sense of pure presence in the intangible present moment. This, John Fowles tells us, "is nature's consolation, its message." Nature "can be known and entered only by each, and in its now; not by you, through me, by any you through any me; only by you through yourself, or me through myself."[23]

Fowles is a writer, but for him words miss this quality of nature. "Such inturned peace, such profound harmlessness, otherness, selflessness, such unusing . . . all words miss," he says; "I know I cannot describe it." Perhaps the words that come closest, Fowles says, are this phrase from another context: "the strange phosphorescence of life, nameless under an old misappellation."[24] To be nameless is to *be* rather than to *say*. The "namelessness" of nature is the guarantor of its value, its ability to insist on our direct presence to it.

Being, not saying: Though it helps us to a sense of presence, entering the silence requires leaving something behind. The students who stayed for long in the Australian rainforest found that human chatter seemed useless. But in order to be willing to experience the difference between chatter—the talk that fills up space but falls away before Cézanne's paintings—and silence, we must have a sense of self that does not desert us in the silence, a sense of self that can be returned to, that allows us to move into an unformed space and then back to a more integrated state.

To enter this space, we must have the capacity to be alone. And this capacity depends in turn on the experience of being both separate and merged with another. It originates, Donald Winnicott tells us, in the initial bond with the mother. We are first alone in the presence of another, the mother, and it is this experience from which we develop the capacity for solitude.[25] Our sense of union with the mother made it safe for us to enter the empty unformed space of play. We were both still enough at one with her and newly separate enough to leave her while in her presence.

To enter this silence of nature that we enter alone may mean to suffer. And so the capacity to experience and suffer pain—perhaps originally the pain of the mother's absence—is also necessary. The woman who cried about the poisoned river in the movie not only had to have enough sense of a bond with her friends that she could leave them and enter solitude; she also had to be willing to open herself to deep grief about the poisoned waters. And my friend who as a boy heard the catfish scream became one with the anguish of a suffering creature at that moment, no matter whether the sufferer was himself or the fish. At that moment for him there was no difference between a human and a fish.

In this suffering we recognize the experience of identification with other life. Indeed, whether encountered through suffering or not, the deep receptivity to the nonhuman world that I am calling "entering the silence" is experienced as identification, as a merging with other life. Is this merging an expansion of the self, or a loss of the self in selflessness? It makes no difference, I think. It is both. In essence, this merging takes us into the direct presence of self and other—and we experience the other from the inside of life as such.

In this profound experience of self and other, merged and yet separately present, silence becomes the fertile void the Buddhists speak of: the void that generates the "ten thousand things," the whole array of created beings. The roaring of the ocean waves arises out of the silence of the ocean. This is the interplay between emptiness and form. Colette Richard, the mountain climber cited earlier, tells us how the silence of the mountain—hardest of hard matter—was alive: "I was there listening with my whole being, and with my whole strength contemplating that mountain that I so dearly love." "Was there anyone in the world, at that moment, as happy as I?" she asks. "For the silence was not emptiness. The silence was Life, making one with the Word. That region was filled with silence—that is to say, filled with life."[26] Chinese nature poems, Gary Snyder tells us, similarly embody this interplay of silence and the essence of life. These poems, he tells us, are not really landscape poems: "Mountains and rivers [are] seen to be the visible expression of cosmic principles; the cosmic principles go back into silence, non-being, emptiness."[27]

What we identify or merge with in the silence of the fertile void is life—livingness—itself. While retaining an awareness of difference, we experience and become one with the life that unites all things. As Kathleen Raine reports, her mother saw that "the moor was alive." Raine tells us how she herself was sitting at her writing table one evening, a fire burning on the hearth:

> All was stilled. I was looking at the hyacinth, and as I gazed at the form of its petals and the strength of their curve as they open and curl back to reveal the mysterious flower-centres with their anthers and eye-like hearts, abruptly I found that I was no longer looking at it, but *was* it; a distinct, indescribable, but in no way vague, still less emotional, shift of consciousness into the plant itself. Or rather I and the plant were one and indistinguishable; as if the plant were a part of my consciousness. I dared scarcely to breathe, held in a kind of fine attention in which I could sense the very flow of life in the cells. I was not perceiving the flower but living it.[28]

The sense of oneness in the life, or livingness, that Raine experienced is profound. "Living form," she called it, combining structure and movement: "a slow flow or circulation of a vital current of liquid light of the utmost purity," spiritual rather than material, "or of a finer matter, or of matter itself perceived as spirit." To this perception of livingness, *all* beings and all life forms are alive. Though the natural world knows death, it cannot be divided into living and nonliving. "Living form," Raine discovers, does not distinguish animal from plant from mineral. A river is alive in this sense, as is a landscape: "Either everything is, in this sense, living, or nothing is."[29]

Though Raine's experience was one of identification with life itself, she retained her awareness of herself throughout. Just as when the fertile void gives rise to the ten thousand things, all are different but all share in the interplay of emptiness and form, so here awareness of difference coexists with awareness of merger or identification. This seems to be a fundamental characteristic of the experience of entering the silence. Snyder puts it well. In spiritual terms, he tells us, the quest for the archetypal wilderness "requires embracing the other as oneself and stepping across the line—not 'becoming one or mixing things up but holding the sameness and difference delicately in mind.'"[30]

Many would call Raine's experience mystical, and she herself notes that she never had this experience to the same degree again. It is out of the ordinary. But the "ordinary" that the experience took her out of was a lack of connection with the deep reality of life itself. "I returned to dull common consciousness with a sense of diminution," she writes, continuing that "I had never before experienced the like . . . and yet it seemed at the time not strange but infinitely familiar, as if I were experiencing at last things as they are, was where I belonged, where, in some sense, I had always been and would always be. That almost continuous sense of exile and incompleteness of experience which is, I suppose, the average human state, was gone like a film from sight."[31] When we enter the silence, we return from the exile that is our ordinary state of mind. To become receptive to the natural world in this way is to come home.

Shockingly Beautiful in Bloom: Wildness, Internal and External

> *The return to the green chaos, the deep forest and refuge of the unconscious . . .*
> —John Fowles, *The Tree*

To enter the silence is to touch the fertile void and the nature of life itself. Inner and outer meet at the threshold of that silence. To enter it is to enter the "beyond" of both. For Laurens van der Post, the journey to the interior of the desert is equally the journey to the interior of the psyche. For Basho, the yearlong walk is the road to the interior.[32] Let us call the "beyond within" our unconscious, and the "beyond outside" the wilderness. Both are essentially wild, and both feel alien and frightening to the everyday self. But when we enter the void that is beyond, we feel the oneness; we feel that we have reached our essential natures. "Poets don't sing about society," Gary Snyder reminds us; "they sing about nature. . . . Class-structured civilized society is a kind of mass ego. To transcend the ego is to go beyond society as well."[33]

"The interior," both the unconscious and the wilderness are called. What is it that links them in this way? "There is something in the nature of nature, in its presentness, its seeming transience, its creative ferment and hidden potential, that corresponds very closely with the wild, or green man in our psyches," John Fowles writes.[34] We can meet nature in its creative vitality only with our direct presence, only by opening our interior selves to it. Snyder explains it by distinguishing "nature" as the subject of science from "the wild": "Nature can be deeply probed, as in microbiology. The wild is not to be made subject or object in this manner; to be approached it must be admitted from within, as a quality intrinsic to who we are."[35] To know nature as the wild is not to know external things with the exterior of ourselves but to meet nature's open interior with the interior of ourselves, at the point where the interior as such is the quality of wildness.

This journey to the interior means an encounter with something other, and thus unknown, in ourselves as well as outside ourselves. "Yet Being Someone Other," van der Post titled his memoir. In some sense this journey to the interior means coming home, but it begins with leaving home, as in fairy tales. The quest that ensues is arduous and dangerous. "You may never see home again," Snyder warns us. "Loneliness is your bread. . . . Untied. Unstuck. Crazy for a while."[36]

Here too we can expect silence and speechlessness. To become unstuck may mean losing speech, the speech of home. To encounter the strange, the weird, is, as Michael Ian Paul points out, to encounter that which is in the process of becoming, of coming to consciousness.[37] It will have no words yet. To journey into the wild, the wilderness without and within, is to meet the self-willed, the god within; to experience awe. It does not yield its meaning immediately, and when first there we will have no words. But what emerges from the shock of the wild and the weird may surprise us with its beauty: "Culture is an orchard apple; Nature is a crab," a farmer friend told John Muir. "To go back to the wild," Snyder comments, "is to become sour, astringent, crabbed. Unfertilized, unpruned, tough, resilient, and every spring *shockingly* beautiful in bloom."[38]

Ah, But Nature Speaks to Us After All

> . . . *a language far more powerful and no less exact; a language in which bird and beast and tree were themselves the words, full of otherwise inexpressible meaning.*
> —Kathleen Raine, *Farewell Happy Fields*

When we open to the depths within us, we become receptive to the language of nature. Nature speaks to us, and we hear her. This is what happened to van der Post and his companions when they fell silent. The birdsong became the expression of essence, and the lion's roar was the voice of a god.

As a poet, Kathleen Raine knows that human language can express an experience of vivid life. She also knows that this world of vivid life speaks in its own language as well. In the Borderlands between Scotland and England where she grew up, she tells us, not only certain places but certain animals and birds as well belonged to that world of vivid life. They spoke in their own language: "The hawk, the heron, and the raven, appearing in all the power of their magical significance, would write across the sky some word of ill-omen or of splendour."[39]

To hear nature speak in this way is not a matter of understanding the specific cries of the animal or bird. Nor is it a matter of mere superstition or of conventional symbolism. Rather, in perceiving the creature, we enter into a confluence with a deeper mystery in which both of us participate: "To see the heron, or the crow, was," Raine continues, "in that remote world, an epiphany of indescribable mystery; not the mystery merely of the bird itself (though that was part of it) but of something which concerned also the seer of the bird, a confluence of our existence with the existence of the bird and of both with something else otherwise."[40]

They who lived in those lands felt, Raine writes, that the bird itself knew what it was communicating. The "corbie," the crow, communicated evil, and was an evil bird. But she is not satisfied with her explanation: "To put it so is to translate—and how inadequately—into the language of words what was itself a language far more powerful and no less exact; a language in which bird and beast and tree were themselves the words, full of otherwise inexpressible meaning. Poetry was not then words on the page, but birds in the air, in the dusk, against the wind in the high blue air; it was trees, it was stones and springs, an ever-changing face of things which communicated knowledge words can only remotely capture or evoke."[41]

"Bird and beast and tree were themselves the words": the language of poetry may eventually express something of this otherwise inexpressible meaning, but as we saw with van der Post, one's initial experience is wordless. We may first hear the voice of the natural world in the form of dreams, visions, songs— or simply impulses, a sense of being influenced. Snyder quotes a Crow Elder: "You know, I think if people stay somewhere long enough—even white people—the spirits will begin to speak to them. It's the power of the spirits coming up from the land. The spirits and the old powers aren't lost, they just need people to be around long enough and the spirits will begin to influence them."[42]

When humans stay long enough, they open themselves to the notice of the earth and its spirits. They hear the voice of nature because nature has noticed them. Entering the silence thus means becoming part of a reciprocal perception in which one both knows and is known. Van der Post speaks of a Great Memory, the larger memory of the earth that remembers us, in distinction to our small memories of ordinary events, and tells of his feeling that the Bushmen living in the Kalahari had both a sense of belonging and a sense of being known. And is not being truly known the essence of belonging?

Culture and the Wild

Poets don't sing about society. They sing about nature. . . . Class-structured civilized society is a kind of mass ego. To transcend the ego is to go beyond society as well.
—Gary Snyder, "Poetry and the Primitive"

Yes, I would like to hear the birds there in the sea. I would like to hear those ocean birds sing in my native language.
—Papago elder

To enter the silence is, as Snyder says, to transcend society. But at the same time culture, our social home, finds ways to help us enter the silence. It shows us what we need to do to continue to be part of the reciprocity of perception.

Culture provides, we hope, a template for the adult counterpart of Whitman's child who goes forth and becomes a part of all that he encounters. The adult must earn the right to continue being part, to be noticed and remembered. The Bushmen dance as long as the moon is up to show the moon that they care about it. The Papago hunter tries to make himself worthy of receiving songs in dreams. Gary Paul Nabhan suggests apprenticing ourselves to other species to learn their stories. The role of culture is to provide opportunities for such apprenticeships, to offer receptacles for the repertory of stories and occasions for their renewal, to remind us to try to become worthy of songs and attend to our dreams. Such cultures, which maintain the link with the interior, might be considered "wild societies."

The adult needs both human cultural expression and the direct experience of nature's own speech. Nabhan tells of a visit to an elderly Papago friend who expresses the wish to go to the ocean (the Sea of Cortez, as we call the place he meant) to hear the ocean birds sing before he dies. In the old days the people went there, the man explains, and when they came home the ocean birds began singing in their dreams, and the songs showed up in the music and dance of the group. Now he wants to hear those birds himself. "In the end," Nabhan reflects, his friend "sought to juxtapose his culture's aural imagery of ocean birds with what the birds themselves were saying. He desired to experience nature directly, as a measure of the cultural symbols and sounds he had carried with him most of his life."[43]

In the absence of a culture to foster our intimacy with wildness, we are left speechless—not only deaf but dumb. This same elderly Papago man, Nabhan tells us, brought home a mockingbird for his mute son to eat in order to acquire the bird's loquaciousness. Inner and outer are in dynamic relation to one another, Nabhan comments. They cannot be separated. We take in, we express out. There is much of the outer, of other species, literally in and on every individual body. But whereas all these other creatures have stories of their ways of living, we are pitifully limited in the stories we hear. "I can't help but wonder," he continues, "if the dilemma of our society is not unlike that of the mute child who needs to eat the songbird in order to speak. Unless we come to embody the songs from the Far Outside, we will be left dumb before an increasingly frightening world."[44] Here speechlessness—being left dumb—means having no way to hear and learn the stories of the natural world. Without the means to make the journey to the interior—or the far ocean, as the case may be—to receive nature's stories, without the means to enter the confluence and experience oneness with life, the world without becomes alien and frightening instead.

The Wildness of Language in Turn

Consciousness, mind, imagination and language are fundamentally wild. "Wild" as in wild ecosystems—richly interconnected, interdependent, and incredibly complex. Diverse, ancient, and full of information. . . .
—Gary Snyder, "Unnatural Writing"

If cultures are wild societies, human languages are wild as well. Snyder is the spokesperson for the wildness of language. Languages are not human inventions so much as naturally evolved wild systems, he reminds us, and each of the world's four thousand or so languages has its own patterns and syntax, which no one invented.[45] Wildness internal and external is self-willed and self-organizing, generative like the fertile void of which it is an instance. Hence we are wrong to accuse language of reducing and impoverishing the world of experience. To the contrary, language reflects the wild generativity of the mind.

Doug Peacock, author of *Grizzly Years,* was disappointed to find out that Annie Dillard considered herself primarily a *writer.* Apparently he considered writing something interposed between oneself and the natural world. Snyder, however, argues for a writing that would be an "art of the wild." Such an art might be very different from what we know of as "nature writing." Rather than writing words *about* (and therefore distant from) nature, we might let language itself express nature. The "art of the wild" would then mean understanding art in the context of nature as process, because "wild" is a name for the way phenomena actualize themselves, the way they emerge from the fertile void.[46] Language is wild when we let it emerge from the full experience of the present moment: "To see a wren in a bush, call it 'wren,' and go on walking is to have (self-importantly) seen nothing," Snyder asserts. But "to see a bird and stop, watch, feel, forget yourself for a moment, be in the bushy shadows, maybe then feel 'wren'—that is to have joined in a larger moment with the world."[47]

"Language," Snyder tells us, "is woven into the seeing, feeling, touching, and dreaming of the whole mind."[48] In its openness and fertility, in fact, language is inseparable not only from the mind but also from the body, indeed, from the totality of our experiencing. It is not simply that language, as a product of the wild mind, mirrors the wildness of nature outside it. Rather, if "the world is our consciousness, and it surrounds us,"[49] as Snyder says, then we can think of landscape and language as two regions of the same total consciousness. The dialect of the Scottish border country, Kathleen Raine tells us, was generated by the landscape itself, transformed by experience and imagination acting on the qualities of that landscape. "There is no 'in general,'" Stanislavski was always reminding his actors. "The words of a living dialect are not abstract terms," Raine writes, as if echoing him. "They are an imaginative map of a country of the imagination," and of our experiences in that world, "caught up into regions of consciousness, with the peculiar light and climate of that place."[50]

If living words are inseparable from the landscape, words themselves are individuals, compounded of matter, imagination, mind, and body. The words of the "braid Scots" her mother's family spoke, Raine tells us, "are concrete as if the landscape itself had uttered them: 'In behint yon auld fail dyke' can no more be translated as 'close behind that old turf wall' than can 'corbie' as 'crow'; for the words bring with them the context of their world, are inseparable from that total experience." The words of the ballad called "Twa Corbies" evoke the landscape of the Borderlands, where the harsh sounds of the crows are carried over the bare hills "to the ears of those who themselves participate in that total consciousness of which landscape and language are alike regions."[51] Language too participates in that confluence of matter and spirit that is the experience of life as one.

Notes

1. See Theodor W. Adorno, *Aesthetic Theory,* 69: "To feel nature, and most of all its silence, has become a rare privilege. . . . The disinclination to talk about [natural beauty] is strongest where love of it survives. The 'How beautiful!' at the sight of a landscape insults its silence and reduces its beauty. . . ." And Gary Snyder, on respectful silence: "Perhaps one should not talk (or write) too much about the wild world; it may be that it embarrasses other animals to have attention called to them. A sensibility of this sort might help explain why there is so little 'landscape poetry' from the cultures of the old ways. Nature description is a kind of writing that comes with civilization and its habits of collection and classification" ("The Etiquette of Freedom," in *The Practice of the Wild,* 21).

2. Note the saying attributed to Baba Dioum, "We only protect what we love."

3. See Otto Kernberg, *Love Relations,* chap. 12, "The Couple and the Group."

4. Neil Evernden, *The Natural Alien,* 16–17.

5. Susan Griffin, *A Chorus of Stones,* 65.

6. Griffin, *A Chorus of Stones,* 15.

7. Alfred Wolfsohn, *The Problem of Limitations,* quoted in Noah Pikes, *Dark Voices,* 33.

8. Griffin, *A Chorus of Stones,* 33.

9. Griffin, *A Chorus of Stones,* 44.

10. Griffin, *A Chorus of Stones,* 291. On Wolfsohn, see Griffin, *A Chorus of Stones,* 289–299.

11. Alfred Wolfsohn, *Orpheus,* in Pikes, *Dark Voices,* 35.

12. Donald Meltzer, *Dream-Life,* 69.

13. For a readable exposition of Bion's sometimes difficult ideas, see Joan Symington and Neville Symington, *The Clinical Thinking of Wilfred Bion.*

14. She wanted to tell her that she had experienced the moor as alive. See Kathleen Raine, *The Land Unknown,* 120.

15. Lionel Corbett, *The Religious Function of the Psyche,* 127. On suffering and the incarnation of the Self, see all of Corbett, chap. 7, "A Depth-Psychological Approach to the Problem of Suffering."

16. See Laurens van der Post, *About Blady,* 76: "In this ancient and beautiful Bantu metaphor, each tear of sorrow shed is a jewel that has to be strung on an unbroken thread of feeling into a necklace, which one can hereafter wear as an ornament of grace around the image of one's spirit, and so prepare the way for the final metaphor, 'Let in our sister, Grief, who should always have a place by our fire.'"

17. Maurice Merleau-Ponty, "Cézanne's Doubt," in *Sense and Non-Sense,* 16.

18. In Lucinda Vardey, ed., *God in All Worlds,* 523.

19. In Michael Murphy and Rhea White, *In the Zone,* 12.

20. Laurens van der Post, *The Voice of the Thunder,* 82–83.

21. Van der Post, *The Voice of the Thunder,* 82–83.

22. Van der Post, *Yet Being Someone Other,* 147.

23. John Fowles and Frank Horvat, *The Tree* (unpaginated).

24. Fowles and Horvat, *The Tree* (unpaginated).

25. See D. W. Winnicott, on "The Capacity to Be Alone," in *The Maturational Processes and the Facilitating Environment,* 29–36.

26. In Michael Murphy and Rhea White, *In the Zone,* 12.

27. Gary Snyder, "The Great Clod Project," in the *Gary Snyder Reader,* 293–294.

28. Kathleen Raine, *The Land Unknown,* 119.

29. Raine, *The Land Unknown,* 119.

30. Snyder, "Survival and Sacrament," in *The Practice of the Wild,* 180.

31. Raine, *The Land Unknown,* 120.

32. This theme runs through van der Post's works on Southern Africa, as for instance in the title *Venture to the Interior.* The title of Basho's diary of his journey is sometimes translated as *Narrow Road to the Interior.*

33. Snyder, "Poetry and the Primitive," in *Earth House Hold,* 122.

34. Fowles and Horvat, *The Tree* (unpaginated).

35. Snyder, "Survival and Sacrament," in *The Practice of the Wild,* 181.

36. Snyder, "Survival and Sacrament," in *The Practice of the Wild,* 179–180.

37. See Michael Ian Paul, *Before We Were Young,* chap. 2, "The Sense of Strangeness."

38. Snyder, "Survival and Sacrament," in *The Practice of the WIld,* 179.

39. Kathleen Raine, *Farewell Happy Fields,* 24.

40. Raine, *Farewell Happy Fields,* 24.

41. Raine, *Farewell Happy Fields,* 25.

42. Gary Snyder, "The Place, the Region, and the Commons," in *The Practice of the Wild,* 39.

43. Gary Paul Nabhan, "The Far Outside," in David Clarke Burks, ed., *Place of the Wild,* 23.

44. Nabhan, "The Far Outside," in David Clarke Burks, ed., *Place of the Wild,* 27.

45. Snyder, "Language Goes Two Ways," in *A Place in Space,* 174.

46. Snyder, "Unnatural Writing," in *A Place in Space,* 168.

47. Snyder, "Language Goes Two Ways," in *A Place in Space,* 176.

48. Snyder, "Language Goes Two Ways," in *A Place in Space,* 175.

49. Snyder, "The Etiquette of Freedom," in *The Practice of the Wild,* 16.

50. Raine, *Farewell Happy Fields,* 27.

51. Raine, *Farewell Happy Fields,* 27.

2   The Love of Nature and the Concern for Life

. . . holding the sameness and difference delicately in mind.
—Gary Snyder, "Survival and Sacrament"

In the beginning we are not separate. Our bodies are made of precisely the same stuff as the rest of the world, organic and inorganic. Only as we grow and develop as human beings do we become more separate in our selves. At first we are aware of the continuity and similarity of ourselves with the rest of nature, but eventually we come to emphasize the difference in order to sense who *we* are, separate from the world we love. But the fact of relatedness never goes away. As we become more mature, we grow the capacity to acknowledge and experience our kinship with what is most different from us, "holding the sameness and difference delicately in mind."[1]

The feelings of attachment and identification we develop early in our lives—the feelings that give rise to our sense of "home"—become the basis of our later recognition of kinship. We know attachment and identification because we have experienced it; it is the matrix from which the self develops. But it is only by suffering the loss of those early bonds—and enduring the pain and destructiveness that were in any case bound up with the love in them—that we become able to feel concerned for what we love, a concern that includes the knowledge that we are also separate and responsible.

The natural world is part of the early home to which we bond and in which we form our selves, but it is also the unfamiliar world from which we distance ourselves. Some retain those early bonds, while some discover the natural world as though it were a wholly different world. I think it is something that, in maturing, we discover, and rediscover, both as self-like—kin—and as an object of our concern.[2]

Abiding Loveliness: The Place of the Ancestors

What we called "wilderness" was to the Indian a homeland, "abiding loveliness" in Salish or Piegan.
—Doug Peacock, *Grizzly Years*

Originally the womb provides the experience of a sensuous protective enclosure continuous with growth and activity. After birth, the relationship with the mother is the safe and fruitful enclosure. Soon it is the enclosure of the home environment and the sense of family. But home and family are located in Place—what we have come to call an "environment." An environment is something total, a world unto itself, that is, enclosed. And that enclosure, that "surround," is intimate and close with the feeling that the things of that place are "friendly," that is, usable. Living within that environmental enclosure, we are practiced by the things there as much as we use those things in the most practical of ways.

The enclosed world of "home" extends back through family to the ancestors. Home is a place saturated with the ancient continuity of the human life lived there. The place holds that continuous life. Kathleen Raine talks about the sense of the clan within her mother's Scotch family, held together by a common belonging to the land: "It was at its roots a blood-kinship, or, what is almost as strong, a kinship through a common love of the land, of the same mountain and river, the same fields, the same prevailing wind and the sun and the stars rising and setting beyond certain abiding hills."[3]

"Abiding loveliness," Doug Peacock tells us, is the translation of the word the Salish or the Piegan used to refer to their homeland—the same land we European-Americans would have disparagingly called wilderness.[4] "Abiding loveliness" speaks both to the sense of continuity over time that the home place provides and also to the ability to see and experience that place as saturated with the love that has circulated within it, a love that flows not only from person to person but between other beings as well.

And so the home place or home*land* for which we feel nostalgic is saturated with a sense of continuity through time. That home place surrounded us with continuity that was embodied in the people we were close to, people who inhabited a place and a time when body and land were still one. Raine speaks to this when she recalls a friend telling her about a Spanish word, *querencia,* which means "love of that specific kind which draws us back to that place from which we have come; as old people as they approach death long to return to their native place; or as a mortally wounded bull in the bull-ring will attempt to walk in the direction of the field from which it has been brought."[5]

If home is the land that surrounds us, saturated with family and ancestors, then home is something we are inside, something that encompasses us. The earth itself is home in this sense. David Abram talks about the "enveloping" or the "encompassing" earth.[6] We do not live *on* the earth, he says; we dwell *within* it, as we realize when we imagine looking down on the earth's atmosphere from beyond the earth, ourselves within it. Abstract Newtonian notions of time and space act to separate and disengage us from this enveloping earth and to disengage time from space. But if we reflect on time and space as experienced, we see that the earth provides space, and we live within that spaciousness as our present. It is our ground. In the same way, Abram suggests, the past is what is concealed beneath the ground, the mythic time in which native peoples like the Hopi emerged, and the future is what is hidden beyond the horizon. In this way the earth, which provides both ground and horizon, becomes the source and origin, the home within which we live our lives toward an unknown future.[7]

The World of Childhood and the Sense of Place

The horizon's edge, the flying sea-crow
the fragrance of salt marsh and shore mud,
These became part of the child who went forth every day.
—Walt Whitman, "There Was a Child Went Forth"

There is a sense in which each creature possesses its world far more real than the mere token ownership conveyed by deed of property. "That all the world is yours, your very senses and the inclination of your mind declare."
—Kathleen Raine, *Farewell Happy Fields*

Nature enters our experience in childhood in the form of place. This place is not so much made up of individual beings—though it is often that too—but is the "world" of childhood, a world of people, dwellings, fields, birds, streams, trees, hills, clouds. The world of childhood is a place to be, a place to become a self. And that world becomes us, literally. In being an *environ*, surrounding us and gathering us inside itself, it gets inside us, providing the very ground of our being as a felt sense of interiority.[8]

This childhood world is vivid, and it is largely lost to us as adults. Though it contains bits of language embedded in it, and the learning of language, and though it may be recreated in language later in life, that childhood world is not primarily a world of language. Its vividness and our loss of it are connected with the fact that it is not a world of words. We take the world of childhood in through all our senses, as a place that contains smells, textures, warmth and coolness, as well as sights and sounds, all together inside the environment that surrounds us. Here, for instance, is John Steinbeck describing childhood memories of the Salinas Valley in Central California: "I remember my childhood names for grasses and secret flowers. I remember where a toad may live and what time the birds awaken in the summer—and what trees and seasons smelled like—how people looked and walked and smelled even."[9]

If later on we lose our attention to and our capacity for this synesthetic kind of experience, we are lost from the earth and its odors. We are outside the interior of things and know them only vaguely through memory. The world as a whole, reflecting our sense of environmental enclosure, is lost, and memory seems only to lead us to a doorway through which we catch glimpses of it. For what became memory was not originally experienced in the same "worded" state of mind in which we now live as adults. Hence childhood amnesia, which, as Ernest Schachtel says in his book *Metamorphosis,* is akin to the forgetting of dreams.[10] We lack a state of mind continuous with childhood experience. Nostalgia is the sad awareness of this irrevocable loss, combined with the faint memory of synesthetic world experience.

Giuseppe Tomasi di Lampedusa, author of *The Leopard,* a Sicilian prince and the last of his line, was tremendously nostalgic for his lost childhood homes, one of which had been destroyed in the Allied bombing of Sicily in 1943. In the last few years of his life, at his wife's urging, he began to write about his childhood memories. The first thing he wrote was a memoir titled "Places of My Infancy." There he comments on the excursions of his childhood. He has kept only vague memories of those excursions, he writes, "though the word 'vague' is not quite exact; a better phrase would be 'difficult to describe.' The visual impression has remained vivid in my mind but was not then linked to any word."[11] Memory is itself dependent on the state of mind we are in when it is formed. For Tomasi di Lampedusa, to write his memoir was to acknowledge his loss, and to account for it. After that he was free to write *The Leopard.*

Imagination and Intimacy

> *That people could come into the world in a place they could not at first even name and had never known before; and that out of a nameless and unknown place they could grow and move around in it until its name they knew and called with love, and call it HOME, and put roots there and love others there; so that whenever they left this place they would sing homesick songs about it, and write poems of yearning for it, like a lover . . .*
> —William Goyen, *The House of Breath*

> *Our memories are encumbered with facts.*
> —Gaston Bachelard, *The Poetics of Space*

Memory, like the childhood world itself, is formed of imagination. "Out of a nameless and unknown place they could grow and move around in it until its name they called with love,"[12] writes William Goyen, evoking the intertwining of imagination and intimacy that creates the place that is home. Intimacy grows in a specific place that one loves. The enveloping earth encompasses us when we become particular about a place and stand there long enough to embody it. As Gary Snyder says, "It is not enough just to 'love nature' or to want to 'be in harmony with Gaia.' Our relation to the natural world takes place in a place."[13]

"Intimacy occurs when we live in a world of particular, concrete events, noticeable for what [William James] called their eachness," writes James Hillman.[14] Indeed, as Jack Turner reminds us, the word for the "ten thousand things" that the Buddhists say emerge from the void can also be translated "intimacy."[15] A labor of imagination is needed to call forth the qualities of the particular thing. The aesthetic response does this work, Hillman says, through a kind of inspiration, a breathing-in of the world, in which we help the individual things appear to us in their loveliness by mingling them with our hearts. When we take the particular into our hearts in this way, it responds with its own imagination, revealing "its heart and its soul, becoming personified and thereby lovable—lovable not only to us and because of us, but because its loveliness increases as its sense and its imagination unfold."[16]

Our personal secret places are the ones we remember best from childhood; they are the most intimate. Children seek out refuges and nests, places near the ground and close to home, rather than expansive open vistas.[17] Secret places and places of refuge feel so important to us because the intimate exchange of imagination creates the sense of place. Woods, great fermenters of imagination, are refuges and sanctuaries. We create home and become ourselves in places that not only envelop us but also offer themselves as chambers of imaginative intimacy.

The house offers itself for the creation of home. Tomasi di Lampedusa's "Places of My Infancy" is almost entirely devoted to the ancestral houses of his family. The house that is the "home" of the world of childhood is created of imagination and mingled with the sense of self. The literal house will be dismantled and reorganized in memory: "As I see it now, the way it appeared to my child's eye," writes the narrator of Rilke's *Notebooks of Malte Laurids Brigge,* "it is not a building, but is quite dissolved and distributed inside me."[18] Facts are a burden to memory; if the portion of imagination is removed, the house shrinks and crumbles to dust. As Snyder reminds us, "If on the one hand the house imbued with imagination becomes a sacred place, on the other hand, the house shares in the impermanence of the world as a whole: the 'literal house,' when seen as just another piece of the world, is itself impermanent and composite, a poor 'homeless' thing in its own right . . . made up of the same world as you and me and mice."[19] The house of childhood, as Goyen's *House of Breath* tells us, is formed by the imagination's work of breathing in and breathing out as we live and love within it. In this sense, the earth too, that world we and the house are both part of, is a house of breath.

But intimacy brings with it not only love and loveliness but also hatred and ugliness, pain in many variants. Home is not a place where everyone flourishes. Home may be harsh and ugly, emotionally and physically. The house of childhood may be many houses, none sheltering. Home may be less a place of abiding than a place of constant leaving, exile, less home than homelessness.

Still, imagination makes use of what it can find, and the self finds emotional sheltering when and where it can. For we need a home in order to become a self, and we must make a home as best we can. Seemingly unlikely things can serve as scaffolding for the creation of the world of childhood. A twelve-year-old African-American girl growing up in the city told Robert Coles she did it this way, using not her memory (quite possibly neither she nor her parents, nor indeed her grandparents, had lived anywhere but the city) but her own eyes and what Coles calls "her mind's imaginative search":

> A lot of time . . . I wish I could walk out of that school and find myself a place where there are no whites, no black folk, no people of any kind! I mean, a place where I'd be able to sit still and get my head together; a place where I could walk and walk, and I'd be walking on grass, not cement, with glass and garbage around; a place where there'd be the sky and the sun, and then the moon and all those stars. At night, sometimes, when I get to feeling real low, I'll climb up the stairs to our roof [she lived in a triple-decker building with a flat roof], and I'll look at the sky, and I'll say, hello there, you moon and all your babies—stars! I'm being silly, I know, but up there, I feel I can stop and think about what's happening to me—it's the only place I can, the only place.[20]

A World of Wounds

One of the penalties of an ecological education is that one lives alone in a world of wounds.
—Aldo Leopold

Removed from their home, living things become marginal, and what becomes marginal is diminished or destroyed.
—Jack Turner, *The Abstract Wild*

What the imagination has bound together brings great pain when sundered. Because the natural world is so entwined with our sense of home and our sense of self, we are constantly subject to loss. Leopold felt that the damage he could see with his ecological eye would be invisible to others. In fact, that damage has become more and more visible, to laypeople as well as experts. But Leopold is also speaking to the sense of loss that comes from the destruction of the childhood world, so that the intimacy we knew in and with it lingers only tenuously in memory or as a sense of an elusive possibility.

New losses evoke old ones. By adulthood we have acquired a vast layering of losses. A friend told me of a recurrent nightmare he now has. He dreams that he returns to his grandfather's farm, where he spent happy summers as a boy, only to find it fenced in and destined to become a housing development. If this dream, which is a nightmare, seems banal it is only because it so perfectly demonstrates the way the loss of childhood place and the destruction of the natural environment evoke one another in all of us.

"The ashtree growing in the corner of the garden was felled," wrote Gerard Manley Hopkins in a diary entry. "It was lopped first," he continues: "I heard the sound and looking out and seeing it maimed there came at that moment a great pang and I wanted to die and not see the inscapes of the world destroyed any more."[21] "There came at that moment a great pang": how many of us feel this acute grief at the destruction of trees and forests! In this grief too we see how the current destruction of nature coincides with the loss of the world of childhood. For trees survive in part through their longevity and their abidingness, evoking our sense of the sheltering and maternal. "Perhaps it is this passive, patient nature of their self-preservation," writes John Fowles in his book *The Tree,* "that has allowed man, despite his ancient fears of what they may harbor in terms of other creatures and the supernatural, to forgive them in one aspect, to see something that is also protective, maternal, even womblike in their depths."[22] Maternal, but also defenseless: trees also represent our vulnerability and the vulnerability of our world. They are defenseless against us. "Unlike white sharks, trees do not even possess the ability to defend themselves when attacked," Fowles notes. "They are the most defenseless of creation in regard to man, . . . and so the most prone to destruction."[23]

In his book *Totem Salmon,* Freeman House quotes a saying attributed to Brave Buffalo, a Sioux elder: "I have noticed that all men have a liking for some special animal, tree, plant or spot of earth. If men would pay more attention to these preferences and seek what is best to do to make themselves worthy of that toward which they are so attracted, they might have dreams which would purify their lives."[24] The elder's words suggest a hope: if we pay more attention, make ourselves worthy, we may become purified. Purification may in this case mean a transformation in our sense of relatedness. When we lose the world of childhood, we become more separate. Grief reminds us that we are also related, though no longer enclosed within that world. Mourning loss may be the manifestation of a development within us that becomes concern, a striving to become worthy of what we have loved and to act accordingly.

How does one make oneself worthy of something that has been destroyed? The loss of the sense of place that shaped the childhood world, the homeland, can lead to terrible despair. One of the symptoms of acute grief is guilt. Somehow, we feel, we are responsible for the death, the destruction; somehow we could and ought to have prevented it. The balance between guilt and purification through mourning is delicate. Loss that cannot be turned to creative use may be experienced as overwhelming guilt. Guilty, we feel ourselves unworthy. Feeling ourselves unworthy, we may fail the things we love. Rather than purifying ourselves by striving to become worthy, we may try to expurgate our guilt through further destruction. We may punish ourselves by multiplying our losses, essentially obliterating ourselves.

"Self-dispossession," writes Erich Lindemann, the psychiatrist who studied grief in survivors of the devastating Coconut Grove fire in Boston in the 1940s, is one of the ways acute unresolved grief can be manifested. Patients engaging in self-dispossession, Lindemann says, "with un-called for generosity, give away their belongings, are easily lured into foolish economic dealings, lose their friends and professional standing by a series of 'stupid acts,' and find themselves finally without family, friends, social status, or money. This protracted self-punitive behavior seems to occur without any awareness of excessive feelings of guilt. It is a particularly distressing grief picture because it is likely to hurt other members of the family and drag down friends and business associates."[25] "I wanted to die," wrote Hopkins—the ultimate form of self-dispossession. But he refrained, and wrote that in his diary instead.

Hard Matter in Its Home: What Wildness Preserves

> *"The* solid *earth! The* actual *world!"*
> —Henry David Thoreau, *The Maine Woods*

> *Thoreau's famous saying, "In Wildness is the Preservation of the World,"*
> *asserts that wildness preserves, not that we must preserve wildness.*
> —Jack Turner, "The Quality of Wildness"

The Native American vision quest was necessarily performed in the wilderness, in solitude, away from home. The Pikuni of the Badger Two Medicine Wildlands, for instance, believed that dreaming at home was unlikely to convey the kind of power the vision quest was intended to bring contact with.[26] Are wildness and home opposites, so that one must sometimes leave home and journey forth into the wilderness? Is the opposite of wildness the domestic and the familiar, the comfortable and cozy? Is there another home, unfamiliar to us?

As we become more separate yet more worthy of what we love, concern evolves out of the experience of the loss of the childhood home. In the process we become worthy of another kind of home, the one Thoreau would call the home of hard matter. Here is how he describes his experience on Mt. Katahdin in Maine:

> We walked over it with a certain awe. . . . What is it to be admitted to a museum . . . compared with being shown . . . some hard matter in its home! I stand in awe of my body, this matter to which I am bound has become so strange to me. I fear not spirits, ghosts, of which I am one,— that my body might,—but I fear bodies, I tremble to meet them. What is this Titan that has possession of me? Talk of mysteries!—think of our life in nature,—daily to be shown matter, to come in contact with it,—rocks, trees, wind on our cheeks! The *solid* earth! The *actual* world! The *common sense! Contact! Contact!*[27]

In wildness is the preservation of the world, wrote Thoreau. Hard matter in its home *is* wildness. For Thoreau, writes Jack Turner, "wildness was a given. His task was to touch it and express it."[28] Thoreau's description of his experience on Mt. Katahdin does precisely that.

"I fear bodies, I tremble to meet them," said Thoreau. There is a dimension of the world, but also of ourselves, that is unfamiliar and frightening, mysterious in its reality and its unfamiliarity. It is real but unknown. This is our wildness, the deeper layer that we can come into contact with and be transformed by as we continue in our more cultured and domestic pursuits. The wild outside—wilderness—matches and is matched by the wild inside, the deeper levels of the mind and spirit and body. In a journal of a sojourn in the Alaska wilderness, John Haines paraphrases Goethe's definition of happiness: "We are happy when for everything inside us there is an equivalent something out there"—a world "alive and sentient, responding to the imagination in shapes of terror and joy." "The wilderness is out there," he continues, "quiet under the brief, rose-gray twilight before the sun rises again. But the wilderness is in myself, also, like a durable shadow. I prowl my region of flesh, my forest of blood, muttering and sniffing, turning many times in search of my own best place."[29]

"The *solid* earth! The *actual* world!": Thoreau's phrases evoke solidity as well as mystery. When we leave the familiar we enter mystery. But we also have the opportunity of seeing the actual reality of the real physical world in its presence and in the present. Unless we wake up to the present, where we actually are, Thoreau is saying, we will miss reality. Waking up to hard matter in its home means accepting our separation from the familiar home that was ours. It means accepting our differentness and holding the sameness and difference delicately in mind. Being more separate from the past, having mourned its pain, allows us to relate to unfamiliar hard matter in its home in our present reality. That is what is wild.

Harold Searles tells a story about this. He grew up in the Catskills in upstate New York, but as an adult he lived and worked in Washington, D.C., and the hills of Maryland. To his mind, nothing was as lovely, nothing could possibly be as lovely, as the hills of his childhood. One day, driving into the area of Maryland where he lived, he found himself thinking that he ought to be in the hometown of his childhood. What am I doing here? he asked himself. At that moment he noticed how little of himself he had invested in the place he was living. On another day he was out in the hills of Maryland and someone with him remarked how beautiful those hills were. "That poor man hasn't seen beautiful hills," Searles found himself thinking, having in mind the hills he had grown up in. Then he looked at the hills around him and noticed that they were indeed beautiful. At that moment Searles realized that he had been living "as if" he were still in the surroundings of his childhood, with the result that he had not seen the actual world he was in. A little later, he realized something else: while consciously he had been thinking that his current surroundings were nowhere near as beautiful as those of his childhood, in fact he had brought with him into adulthood all the misery and poverty of his childhood, and what he was seeing when he looked at his current surroundings was really that unmourned childhood misery, clouding his vision.[30] As he left the childhood world a bit more, he became more able to see the hard matter of his current home.

In wildness is the preservation of the world, Thoreau says. What is it that wildness preserves? If wildness is the quality of being present to the mystery of actual reality (the actual world, the solid earth) in the present moment, then wildness preserves the link between ourselves and that solid, actual natural world. Wildness is the unfamiliar opposite of our familiar reality—whose familiarity may in fact be a manifestation of long uncompleted mourning for earlier pain and loss, as Searles's story shows us. As we mature into greater awareness of our separateness, we become conscious of this unfamiliar relatedness. We come to learn where we stand—on the earth—and to know it is the home we live within.

Becoming Worthy: Language and the Labor of Concern

> *He would be a poet . . . who nailed words to their primitive senses, as farmers drive down stakes in the spring, which the frost has heaved; who derived his words as often as he used them—transplanted them to his page with earth adhering to their roots.*
> —Henry David Thoreau, "Walking"

Hopkins suffered when the ashtree in his garden was felled. He wanted to die and not see the "inscapes of the world" destroyed any more. But he did not die. He did not destroy himself through guilt. Instead, we might say, he made himself worthy of his love, and his writing is both his labor and his reward.

Concern is the effort to make oneself worthy of what one loves. It is expressed in working to know and appreciate the beloved and to preserve its value. Hopkins, like Kathleen Raine, felt that language grew from the land and expressed its essence. His concern expressed itself in intense efforts to learn language specific to place. These investigations are recorded in his diary, as in this entry from 1871: "Robert says the first grass from the scythe is the wathe, then comes the strow (tedding), then rowing, then the footcocks, then breaking, then the hubrows, which are gathered into hubs, then sometimes another break and turning, then rickles, the biggest of all the cocks, which are run together into placks, the shapeless heaps from which the hay is carted."[31] Hopkins's investigations of language occur at the deep level we have called wildness. The "inscapes" he speaks of are internal imaginative pictures of the essence—the eachness—of the thing. Interior meets interior, imagination in the service of wildness, not only our imagination but the imagination of the things themselves. The language that grows in a particular region is an expression of its inscape. Hopkins's own use of language to express the inscapes he saw is another. Here is another line from his diary: "ground sheeted with taut tattered streaks of crisp gritty snow."[32] The essence of the particular thing perceived and the essence of the particular words meet and join in the written line. This is nature as writing and writing as nature, nature speaking in the wildness of language.

Concern encompasses the emotional complexity of relatedness in separateness. It develops from the sense of enveloping we first experienced as children, from the attachment we felt to the things of home, and from our capacity to identify our selves with what is not us. Concern depends on the capacity to meet what we love at the level of wildness. Meeting there, we experience both suffering and joy. Hopkins could not perceive the eachness of things without experiencing pain at their destruction: "—O if we but knew what we do/ When we delve or hew/ Hack and rack the growing green!"[33] But his concern is also inseparable from joy in them: "Glory be to God for dappled things—/ For skies of couple-colour as a brindled cow;/ For rose-moles all in stipple upon trout that swim. . . ."[34]

To become concerned with the eachness of things is to meditate on selfhood. This is the core of Hopkins's concern. To live, to be, to express one's essence, to let one's inscape flower in the imaginations of others is to "self":

> As kingfishers catch fire, dragonflies draw flame;
> As tumbled over rim in roundy wells
> Stones ring; . . .
> Each mortal thing does one thing and the same:
> Deals out that being indoors each one dwells;
> Selves—goes itself; *myself* it speaks and spells,
> Crying *What I do is me: for that I came.*[35]

What Is the Self and Where Are Its Boundaries?

The others who are none other than myself
include mountains, rivers and the great earth.
—Dogen Zenji

Some have said they have felt a boat shudder before she struck a rock, or cry
when she beached and the surf poured into her. This is not mysticism, but

*identification; man, building this greatest and most personal of all tools, has in
turn received a boat-shaped mind, and the boat, a man-shaped soul.*
—John Steinbeck and E. F. Ricketts, *Sea of Cortez*

What is the self and where does it reside? Perhaps its boundaries are not fixed.
Are they located at the skin? In the brain or in consciousness? Where does consciousness end? William James, writing in the late nineteenth century, elaborates on this ambiguity:

> The Empirical Self of each of us is all that he is tempted to call by the
> name of me. But it is clear that between what a man calls me and what
> he simply calls mine the line is difficult to draw. We feel and act about
> certain things that are ours very much as we feel and act about ourselves.
> Our fame, our children, the work of our hands, may be as dear to us as
> our bodies are, and arouse the same feelings and the same acts of repri-
> sal if attacked. And our bodies, themselves, are they simply ours, or are
> they us?[36]

What is *mine* is close but still other than me; it is something I possess. What is
me is part of my self; I identify with it. The things we are closest to are experienced as part of the self and not as mere external possessions. James offers a
list of the things we identify with in this way: the body, the clothes, the family,
the home. "Our father and mother, our wife and babes," he says, "are bone of
our bone and flesh of our flesh. When they die, a part of our very selves is
gone. . . . Our home comes next. Its scenes are part of our life; its aspects
awaken the tenderest feelings of affection."[37] The attachment we feel to things
and places, our love of home and all that we feel to be home: this is a form of
identification. We experience those places and beings outside ourselves as ex-
tensions of ourselves, even as parts of ourselves—as much *me* as *mine*.

The things James mentions fall within a narrow sphere of blood kin and personal possessions. The sense of self extends to the home, that he grants, and
we have seen how imagination mingles self and house. But still, James portrays
identification, the sense of me-ness, as quite limited. Could we, who are human beings, identify our selves with things in some sense more distant and
more alien—creatures of the wild, members of other species, life forms like rivers and mountains? If love is a form of identification, could identification be a
basis for concern for the natural world?

Identification with the wild Other, many would say, is precisely what may allow wild things to be saved. The notion of an "extended self" or an "ecological consciousness" speaks to this idea.[38] If empathy and the sense of self can be extended beyond our identification with human groups like the family to the natural world, perhaps we can stop destroying nature. This broader self that identified with the natural world would be an "ecological self"—a self that identifies with greater wholes. We would then defend the larger world of nature just as we would defend members of our own family. I am part of the rainforest protecting itself, says John Seed, echoing this idea.[39]

In his book *The Fallacy of Wildlife Conservation*, Canadian naturalist John Livingston tells about being on a beach on the South China Sea watching a leatherback sea turtle return to the sea after laying her eggs. At first he watches carefully, "in the usual schizoid way of the naturalist," as he calls it, "marveling at it all whilst at the same time paying keen attention to each minute detail and considering the chances of hatching success, hatching survival, turnover, and all the rest of it." But as the turtle moves into the sea, he allows himself to shift into what he calls "free flow": "She—the turtle—remained on the surface for several moments, possibly to catch her breath. In those moments, as gentle ripples broke soundlessly over the gigantic carapace, now wetly silver in the moonlight, my universe was no longer mine alone. It was ours. With the leatherback turtle, I slipped beneath the smooth surface of the sea. . . . As the ceaseless cycle turns, universe without end, life without beginning, pulse without destiny, form without function, love without self, circling among unknown stars, we are once again in touch."[40] He is not speaking this way in order to add to the mystique of "the wildlife experience," Livingston tells us, but rather "to emphasize that when I say that the fate of the sea turtle or the tiger or the gibbon is mine, I mean it. All that is in my universe in not merely mine, it is *me*. And I shall defend myself."[41] This is what is meant by identifying with the wild Other. And how different it is not only from the naturalist-observer's interested and admiring point of view, but also from the rational-pragmatic argument that our survival is linked to the well-being of our natural—that is, physical and biological—environment!

How far can this kind of identification be taken? The boundary between me and mine is not clear. We can think of the self extending along a spectrum, says Arne Naess.[42] From a narrower, more limited sense of self, our sense of what is self can broaden to include other beings. At its greatest extension, the Self becomes part of all beings, and conversely the Self is experienced as containing all other beings. Many religious traditions articulate this idea of complete identification of the Self with other beings. We see it expressed in the *Bhagavad Gita.* We also hear it in Robinson Jeffers's credo: "I believe that the universe is one being, all its parts are different expressions of the same energy, and they are all in communication with each other, therefore parts of one organic whole. . . . The parts change and pass, or die, people and races and rocks and stars; none of them seems to me important in itself, but only the whole. This whole is in all its parts so beautiful, and is felt by me to be so intensely in earnest, that I am compelled to love it, and to think of it as divine."[43] Kathleen Raine experienced this kind of complete identification with her hyacinth; she felt she was at one with its life flow. Many of us have experienced this shift, at least momentarily. Often these moments remain as touchstones, awe inspiring and felt to be deeply significant even if not yet fully articulated and understood. They become the foundation of a passionate concern for the natural world.

Neil Evernden has written two books on humans' relationship to the larger natural world. We alternate between a sense of larger identification with the self and a sense of separateness and otherness, he says. In the first book, he was interested in the extended self, but now he is more interested in the "shock of the other," he told Derrick Jensen in an interview.[44] We have just heard John Livingston's experience of identification with a sea turtle: "All that is in my universe it not merely mine, it is *me*." But Livingston's writing can equally well be considered the opposite. For Evernden, it demonstrates not the expansion but the loss of self: "the shock of encountering a frog or a cat or just listening at a frog pond."

"Those are the kinds of things you find in people who are strongly motivated to speak in defense of the nonhuman," Evernden comments.[45] Naess titled an essay "Identification as a Source of Deep Ecological Attitudes." Identification, the extension of the self in continuity with the other, might equally be thought of as loss or forgetting of the self in the experience of the separate self of the other. Not only do we extend the feeling of self to encompass other creatures, in that extension we also become able to perceive selfhood in those others. Hopkins expressed his concern in an appreciation of the eachness, the self of other beings, kingfishers and dragonflies: "Each mortal thing does one thing and the same: . . . goes itself." As both expansion of self to include the other and loss of self in awareness of the other, identification moves toward the whole. Sense of self and sense of other are bound together in a larger whole that might be called ultimately Self, not merely my self but a larger Self in which I and others participate.

Steadfast Friendliness: Identification as the Source of Concern

*We are obliged . . . to spread the news, painful and bitter though it may be for
some to hear, that all living things on earth are kindred.*
—Edward Abbey

*To be kindred does not mean we should treat animals as our babies. It means
instead a sense of many connections and transformations—us into them, them
into us, and them into each other from the beginning of time. To be kindred
means to share consciously in the stream of life.*
—Paul Shepard, "Searching Out Kindred Spirits"

Identification with other life is a natural and spontaneous impulse, says Arne
Naess. This is why it can serve as the basis for concern for the earth and its
creatures. Identification, and concern based on it, are evoked most strikingly,
perhaps, by suffering. The news that penguins in the Antarctic were at risk
from the DDT in their eggs evoked spontaneous grief and outrage among humans who had no direct personal experience with penguins. It was identification that gave rise to these emotions.[46] We saw a similar identification arise in
the boy who heard the catfish scream.

The idea that suffering evokes identification in us rings true. Think of how
immediately the cry of a baby in distress arouses similar suffering in us. And
think of how, like the early vivisectionists who cut the vocal cords of the animals they cut apart, we may have felt we had to harden our hearts to resist
helping someone we perceive to be suffering. When Hopkins says that at the
moment he saw the ashtree in his garden lopped and felt "a great pang" and
"wanted to die," he is identifying so strongly with the suffering and the destruction, and finding it so unbearable, that he wants to die himself.

Identification evoked through suffering is likely to become more common as
destruction of the environment and extinction of species continues and acceler-
ates, Naess points out;[47] extinction may put us in touch with sorrow where the
loss of individuals has not. ("Death is one thing; an end to birth is something
else," as Soulé and Wilcox put it in their book *Conservation Biology*.[48]) How
are we to bear the suffering that we will experience? "The ending of the lines
of so many creatures with whom we have traveled this far is an occasion of
profound sorrow and grief," writes Gary Snyder.[49] Is the result of broadened

identification to be such pain that we want to sacrifice ourselves and die? Will we engage in the kind of self-dispossession through guilt that Lindemann described? Increasing awareness of destruction is likely to lead to despair and desperate actions as well as intense sorrow and tireless efforts to save what is left, predicts Naess.

What are we to make of the urge to sacrifice? Love and hate can take violent forms, and passionate defense can be destructive or self-destructive. Concern for life must extend to oneself as a living being as well, as Wilfred Bion points out. If we do not respect ourselves as we do other living beings, we have no way to guard against acting out murderous or suicidal impulses.[50] Youthful sacrifice—indeed the kind Hopkins wished for but did not act on—is often not a defense of the other so much as a wish to put an end to unbearable pain. The destructiveness one feels is turned against the self, as in the self-dispossession of overwhelming grief. We can understand this kind of sacrifice as a reflection of the psychological process of splitting, where all the good is separated from all the bad.[51] The good is idealized and the bad is demonized. The bad, it is felt, must be destroyed, wherever it is located and at whatever cost, including destruction of the self.

Similarly, the choice between egoism and altruism, Naess says, reflects a dualistic thinking that does not conceive of identification with the whole. A whole conceived as a "gestalt," a figure, he points out, does not reinforce the duality of subject and object, the observer and the thing observed, but rather emphasizes the linking, the participation in the larger whole.[52] All creatures are in me and I am in them, but we are still separate and different from one another. When we identify with "the whole and all its parts," we are concerned with life both as a whole and in all its parts, each of them a self—and therefore with ourselves as living beings as well. Such identification does not elicit self-destructive sacrifice.

Sacrifice made in compassionate consideration of the situation as a whole, including the self in the smaller sense, in contrast, would have a different quality to it, a different tone, than self-sacrifice reflecting frustration and desperation and helplessness. Compassion and concern are closely linked, and paradoxically, it is compassion that helps us to bear the suffering we experience in our broader identification with other beings. In the Buddhist understanding of compassion, the suffering of other creatures does not shatter us precisely because we also experience the link with other creatures, and we are supported by our participation in that larger whole.

Alienation is the term Naess uses for the opposite of the natural and spontaneous identification that is the foundation for participation and compassion.[53] The term suggests that we make ourselves alien, not only to the natural world we would otherwise feel a part of, but to ourselves as well. We overcome this largely unconscious alienation, he says, in experiences of identity. If alienation makes us alien to ourselves as well as to the natural world, in widening our identifications—that is, in expanding our participation in the larger whole, the Self—we are in effect displaying an allegiance to our true nature. *In suo esse perseverare*[54]—to persevere in oneself or one's own being, acting out of one's own nature or essence, in a phrase from Spinoza that resonates with Hopkins's "selving"—is to broaden our identifications with other beings. As Harold Searles said, relatedness to what is most unlike oneself requires the highest order of maturity.[55] Or as the Chinese Buddhist saying goes, "Easy to reach Nirvana, hard to enter difference."[56]

What should we call this broadening identification that becomes nonviolent concern for life as such and that means being both connected and separate? What kind of relationship is expressed by concern? Kinship, Naess and others suggest. "Fellow feeling," says John Fowles, noting that fellow feeling finally led to the abolition of slavery, and it will be fellow feeling that will ultimately lead to respect for the intrinsic being of animals and trees.[57] A concerned relationship with other species, Naess adds, might best be thought of as friendship as well as kinship. Friendship emphasizes compassion and devotion in contrast to the passionate intensity of a love relationship, which might entail betrayal, disappointment, and rupture. "Steadfast friendliness," he proposes,[58] suggesting respect for difference and a long-term commitment: a concern that abides.

Creativity, Destructiveness, and the Concern for Life and Truth

> *Hunting is a holy occupation, framed in rules and courtesy.*
> —Paul Shepard, "Searching Out Kindred Spirits"

> *If we do eat meat, it is the life, the bounce, the swish, of a great alert being with keen ears and lovely eyes, with foursquare feet and a huge beating heart that we eat, let us not deceive ourselves.*
> —Gary Snyder, "Survival and Sacrament"

Hunting-gathering cultures identify closely with the animals they hunt, but they know they must engage in killing. The act of killing alienates them from the animals they identify so closely with. How can the identification be restored? This, Naess tells us, is the job of culture—in the form of the rituals and ceremonies surrounding the hunt.[59] The creativity that is sedimented in cultural forms is essential to concern, which must deal with out destructiveness as well as our love.

Concern is more complex than identification conceived of as a pure merging of self and other. It is the work of maturity and of the culture as well as the individual. "Though it is commonly said that primitive men see themselves as 'part of nature,'" writes Shepard, "Paleolithic hunters did not and modern hunters do not 'identify' themselves in a sense of merger or union with nature. The main thrust of their education and culture is the opposite: to clarify and emphasize their identity—not as an apartness so much as the precise definition of sets of relationships.[60]

Concern develops out of the original love and attachment that are early forms of identification, but it is more and other than love. It suffers with suffering and acknowledges destructiveness as well as creativity. Ultimately it is a concern for life as such, and it is based on an allegiance to life as such. "Concern for life does not mean only a wish not to kill," writes Wilfred Bion, "though it does mean that. It means concern for [something] precisely because [it] has the quality of being alive. . . . It means being curious about the qualities that go to make up what we know as life, and to have a desire to understand them."[61]

Bion says that concern for life means being curious about what life is and trying to understand it. Concern for life, he would say, is inseparable from a concern for truth—not the detached observer's attention to mere facts, but the whole of the emotional truth we derive through our capacity to identify with other beings and to be aware of ourselves. Part of that truth is the truth of our destructiveness. We cannot attain concern for life without acknowledging our own destructiveness.

"Let us not deceive ourselves," says Gary Snyder. If we eat meat, "it is the life, the bounce, the swish, of a great alert being with . . . a huge beating heart that we eat."[62] When we eat meat, it is the result of killing, of an alienation that culture must set right again. The fact that we must take life in order to survive, no matter what we eat, epitomizes the unavoidable truth that we all feel and act destructively as well as constructively. Our destructiveness is not simply a matter of killing to eat. It is also that we satisfy our own needs without regard for others. And it is also our ambivalence; we both love and hate our fellow creatures.

The capacity for concern takes shape when the child becomes able to conceive of creative action, says Donald Winnicott, and the ability to be destructive is necessary in order to be creative.[63] We cannot be concerned if we are not capable of feeling guilty for our destructiveness, but it is because we know we are capable of what he calls "contributing in"—that is, helping out with constructive activity—that we are not overwhelmed by our guilt. Winnicott gives the example of an adolescent girl who began to express a good deal of hate and destructiveness toward her therapist in sessions. But at home the girl was spontaneously beginning to be helpful around the house. Her constructive and creative experiences were allowing her to get in touch with her destructive feelings, Winnicott comments, just as conversely her ability to acknowledge and tolerate her hate allowed her to experience her constructive impulses as well. Another patient became able to express his feeling that his therapy with Winnicott had real value. At that point he also became able to express his envy of Winnicott.[64] The valuing was real, as was the envy. The patient was not overwhelmed with guilt about his envy because he could also express his appreciation. He was able to experience and tolerate both feelings in himself.

The capacity for concern first develops in a child's relationship to the mother. The child needs the mother both for survival and as the object of his or her emotional needs. The child wants to "use" the mother in various ways but is dependent on her and begins to realize that using her ruthlessly for personal ends might endanger the child's own survival. This presents a dilemma, and a potential impasse arises in which the child could become unable to act because of being too fearful of the consequences. What makes the difference is the mother's handling of the child's ruthless destructiveness. If the mother does not retaliate, the child will feel that she has "survived" in the sense that she can still be depended on. If the mother accepts the child's efforts to make up for the damage by "contributing in," as Winnicott calls it, the child will not be overwhelmed by guilt. The mother will appear to be safe and the child's needs can continue to be met. The child will not have to divide the world up between people or things used to satisfy desires and those depended on to fulfill needs.

In the adult, concern means among other things tolerating the fact that we often need to kill in order to eat. It means finding ways to renew bonds with those toward whom we feel both hatred and love. It means finding ways to contribute to the natural world on which we depend but that we also use ruthlessly for our own ends. Concern depends on cultural forms for its expression, and it shapes culture in turn, for culture is above all a structuring of relatedness. If we think of the hunting-gathering culture as the prototype of a culture of concern, then as Shepard says, "the main thrust of their education and culture is . . . to clarify and emphasize their identity—not as an apartness so much as the precise definition of sets of relationships."[65]

Transitional Space: Imagination and Creative Action

> *The transitional area of experience, between internal reality and external reality . . .*
> —Donald Winnicott

> *Cave art . . . was precisely a way of accepting reality. Fidelity to a tradition and a sense of connectedness to a real outside world are related, for both rest on a conviction of order and participation.*
> —Paul Shepard, *The Tender Carnivore and the Sacred Game*

A "substitute environment propelling the individual into a new level or plane of awareness"—this is Paul Shepard's definition of a *transitional object.*[66] Donald Winnicott originally coined that term for things like teddy bears and pieces of blanket sucked until they disintegrate—objects that very young children use to retain a symbolic link to the mother while moving into a space in which they feel they are both separate and not separate.[67] What is an object when seen from the outside is to the child a space—the space in which imagination and creative play arise.

Infants feel they have simultaneously discovered and created the transitional object, just as those who inhabit a region, as Kathleen Raine told us, feel an "ownership of Eden," a sense of discovery combined with a sense of belonging. The transitional object exemplifies the intimacy between the inside of the mind and the external world. In this it is the forerunner not only of play and imaginative creativity, but also of culture as meaningful symbolization. The transitional object in its transitional space provides the foundational experience of linkage that allows the perception of the external world to continue to be accessible through imagination.

The transitional object is not confined to the nursery. Anything in the external world can in fact serve as a transitional object or space, for the original symbiotic identification with the nonhuman environment, which remains in our psyches even as we become more differentiated, is the basis on which a transitional object is possible. The house, as we have seen, becomes a transitional object, a nursery of the self, when it becomes home, a creation of the imagination. Secret places are created the same way and serve the same functions.

"*Objets trouvés,*" or "found objects"—the stones, shells, leaves, and feathers we pick up in the outdoors and bring home to set on our windowsills or mantelpieces—function as transitional objects. "Worn, traded, scrutinized, imitated, and treasured," as Shepard notes, they are transitional between humans and nature and between human-made and natural things, the forerunners of works of art.[68] As the teddy bear serves as the substitute for the mother's breast and permits union and separation from it, so the found object continues the link with the natural environment while allowing the human imagination a space in which to work. Found objects are "catalytic in our perception," says Shepard.[69] They combine the feeling of me-ness and the sense of emerging new possibilities that enters into all creative activity.

Imaginative activity prepares the way for concern. Concern means an active concern for life; it gives rise to concerned action. But concerned action is not blind and impulsive, nor can it be conformist. It requires the capacity for creative activity that develops in the transitional space where the baby is both separate and not separate. In the transitional space, all manner of things can be experienced and explored in safety, just as concern in its relatedness can know the passions and destructiveness that are part of life as a whole.

What Naess calls alienation, Neville Symington calls narcissism, and he links it to the failure to develop the capacity for concern and for creative activity.[70] For Symington, narcissism is the essence of all mental disturbance. Narcissism gets set up in the self when one turns away from what he calls "the Lifegiver," the source of nourishment or fulfillment that is necessarily originally outside the self. The person decides to "go it alone" because it is too painful to acknowledge one's dependency on others. Narcissistic structures get built as the result of trauma—situations in which extreme emotional pain is coupled with a lack of the empathic, thoughtful attention that would make the pain bearable. In those circumstances, awareness of need becomes too painful, and the person denies any need for the outside. But in turning away in this way, the self becomes incapable of creative emotional action and relies instead on a surface imitation of others. The depths of others are closed to him.

The alienation that Symington calls narcissism can be understood as a lack of imaginative intimacy with the external world and with the interior world as well. It means that the transitional space cannot be used to give rise to creative play. In adults, or in the culture as a whole, it means that the mature form of creative emotional action, what Symington calls "creative fashioning of the social environment," the creative, life-giving impact we have on the people around us and the cultural environment as a whole, is not possible. Culture's role is to help prevent this alienation and to restore identification and concern where alienation has set in. This is what the rites of the hunting-gathering culture are designed to do for the individual and the group who have had to participate in killing.

In hunting-gathering cultures, Shepard argues, adolescent initiation plays a key role in the development of relatedness to the natural world. (And of course he is speaking in ideal-typical terms, in abstraction from the specifics of individuals, cultures, and historical periods, all of which might fail to achieve this relatedness in specific instances.) In such cultures, it is not the child who has achieved a full, real relationship with the natural world but rather the post-adolescent, the achieved adult. In the adolescent rite of passage, the natural world as a whole serves as transitional object, a "substitute environment propelling the individual into a new level or plane of awareness." In this transitional space the adult-to-be experiences, at least briefly, "full awareness of the cosmic 'fitting together.'"[71] The initiation establishes the natural world, the cosmos, as something with meaning and organization, something that can therefore be related to through culture as an ordering and meaning-making activity. Love of the natural world and regard for tradition thus go hand in hand. They are correlative parts of the whole. As Shepard says, "Fidelity to a tradition and a sense of connectedness to a real outside world are related, for both rest on a conviction of order and participation."[72] Again I am reminded of Robinson Jeffers's statement: "The whole is in all its parts so beautiful, and is felt by me to be so intensely in earnest, that I am compelled to love it, and to think of it as divine." When the natural world becomes a transitional object we are able to perceive the whole as "intensely in earnest." It both shelters us and invites our contributing concern.

Notes

1. This is Harold Searles's argument in his book *The Nonhuman Environment in Normal Development and in Schizophrenia.* See especially the introduction and Part II, "The Nonhuman Environment as Experienced by the Healthy Individual." See also Ernest Schachtel, *Metamorphosis,* on what he calls "the conflict of emergence from embeddedness," 6. The quotation is from Gary Snyder, "Survival and Sacrament," in *The Practice of the Wild,* 180.

2. There is of course tremendous individual variation, not to mention cultural and historical variation, in the makeup of "home," and in the scope, timing, and duration of identifications with aspects of the natural world, as Steven Holmes argues in his discussion of an object-relations approach to environmental biography. See Holmes, *The Young John Muir,* 265–287.

3. Kathleen Raine, *Farewell Happy Fields,* 26.

4. Doug Peacock, *Grizzly Years,* 185.

5. Kathleen Raine, *The Land Unknown,* 116.

6. See David Abram, "Merleau-Ponty and the Voice of the Earth," *Environmental Ethics,* Summer, 1988.

7. Abram, *The Spell of the Sensuous,* 216–217.

8. My friend Janet Gerard speaks of recovering this sense as she becomes more deeply engaged in the life of her current home place in Northern New Mexico: "At one time in my life, I believe I would have said, 'I love nature.' Those were the days (most of my life) when I lived in cities and was separate from nature within my daily routine. I 'loved' to go to the country; and I 'loved' to go hiking. . . . I have begun to perceive nature with a different emphasis. It is no longer separate from me, but rather a part of my routine. . . .When I am here . . . I feel caught up in the rhythm of what nature is; it's patterned into my life" (personal communication, 1999).

9. John Steinbeck, *East of Eden,* 3.

10. See Ernest Schachtel, "On Memory and Childhood Amnesia," in Schachtel, *Metamorphosis,* 279–322. Schachtel paints a picture of the development of perception and the relationship of the self to the object of perception. Infancy, he says, is characterized by what he calls "primary autocentrism," an initial stage in which the self is central. Through the use especially of the proximal senses (taste, touch, smell, proprioception, and so on) the object is felt to be continuous with the self; it is all one world, without a sense of distance or alienation. This initial stage is followed by the stage of childhood play and exploration, when the whole world seems to open up and become full of interest and fascination. The kind of perception governing this stage Schachtel calls "allocentric"—other-centered. External people, things, and places are the object of intense interest in their own right; they open up in their depth to the fascinated child. To the extent that the capacity for allocentric perception is not maintained, however, to the extent that perception begins to follow conventional schemata, perception becomes characterized by what Schachtel calls "secondary autocentrism," in which

the object is assimilated to the preexisting schemata of perception, to the familiar. The newness of the object in its own right, with its inexhaustible depths, is excluded from the perceptual universe.

11. Giuseppe Tomasi di Lampedusa, *The Leopard,* 253.

12. William Goyen, *The House of Breath,* quoted in Gaston Bachelard, *The Poetics of Space,* 58.

13. Gary Snyder, "The Place, the Region, and the Commons," in *The Practice of the Wild,* 39.

14. James Hillman, *The Thought of the Heart and the Soul of the World,* 46.

15. Jack Turner, *The Abstract Wild,* 72.

16. Hillman, *The Thought of the Heart and the Soul of the World,* 48.

17. See Gary Paul Nabhan and Stephen Trimble, *The Geography of Childhood,* 3–8.

18. Rainer Maria Rilke, *The Notebooks of Malte Laurids Brigge,* cited in Gaston Bachelard, *The Poetics of Space,* 57.

19. Snyder, "Blue Mountains Constantly Walking," in *The Practice of the Wild,* 105.

20. Quoted in Nabhan and Trimble, *The Geography of Childhood,* xxii.

21. Gerard Manley Hopkins, diary entry, April 8, 1873, in *A Hopkins Reader,* 111.

22. John Fowles and Frank Horvat, *The Tree* (unpaginated).

23. Fowles and Horvat, *The Tree* (unpaginated).

24. Freeman House, *Totem Salmon,* 54.

25. Erich Lindemann, *Beyond Grief,* 71.

26. Jay Hansford C. Vest, "Sacred Geography of the Pikuni: The Badger Two Medicine Wildlands," in David Clarke Burks, ed., *Place of the Wild,* 87–88, citing Wissler.

27. Quoted in Jack Turner, *The Abstract Wild,* 36–37.

28. Turner, *The Abstract Wild,* 175.

29. John Haines, "Notes from an Interrupted Journal," in David Clarke Burks, ed., *Place of the Wild,* 111.

30. Harold Searles, *The Nonhuman Environment in Normal Development and in Schizophrenia,* 333–337.

31. Hopkins, *A Hopkins Reader,* 105.

32. Hopkins, *A Hopkins Reader,* 110.

33. Hopkins, "Binsey Poplars," *A Hopkins Reader,* 57–58.

34. Hopkins, "Pied Beauty," *A Hopkins Reader,* 50.

35. Hopkins, "When kingfishers catch fire . . . ," *A Hopkins Reader,* 67.

36. William James, *Principles of Psychology,* cited in Arne Naess, "Identification as a Source of Deep Ecological Attitudes," in Michael Tobias, ed., *Deep Ecology,* 259.

37. James, *Principles of Psychology,* cited in Arne Naess, "Identification as a Source of Deep Ecological Attitudes," in Tobias, ed., *Deep Ecology,* 260.

38. For elaborations of the notion of an "ecological self," see for instance Joanna Macy, "The Greening of the Self," in *World as Lover, World as Self,* 183–193, and Warwick Fox, "Transpersonal Ecology and the Varieties of Identification," in Alan Drengson and Yuichi Inoue, eds., *The Deep Ecology Movement,* 136–154. For further elaboration of Naess's concept of the self and self-realization, see Arne Naess, "Self-Realization: An Ecological Approach to Being in the World," in George Sessions, ed., *Deep Ecology for the 21st Century,* 225–239.

39. See Joanna Macy, "The Greening of the Self," in *World as Lover, World as Self,* 184.

40. John Livingston, *The Fallacy of Wildlife Conservation,* 113.

41. Livingston, *The Fallacy of Wildlife Conservation,* 113. Similarly, Rilke recalls a time when "a bird call in the open and in his inner consciousness were one, when it did not as it were break on the barrier of his body, but gathered both together into an undivided space, in which there was only one region of the deepest consciousness, mysteriously protected" (Christopher Bamford, in the Introduction to Jacques Lusseyran, *Against the Pollution of the I,* 10).

42. Naess, "Identification as a Source of Deep Ecological Attitudes," 261.

43. Robinson Jeffers, quoted in Fox, "Transpersonal Ecology and the Varieties of Identification," in Drengson and Inoue, eds., *The Deep Ecology Movement,* 143.

44. Neil Evernden, interview in Derrick Jensen, *Listening to the Land,* 119.

45. Evernden, interview in Jensen, *Listening to the Land,* 119.

46. Naess, "Identification as a Source of Deep Ecological Attitudes," 265.

47. Naess, "Identification as a Source of Deep Ecological Attitudes," 269.

48. Quoted in Gary Snyder, "Survival and Sacrament," in *The Practice of the Wild,* 177.

49. Snyder, "Survival and Sacrament," in *The Practice of the Wild,* 176.

50. Wilfred Bion, *Cogitations,* 247–248.

51. See Elliot Jaques, "Death and the Mid-Life Crisis," in Elizabeth Bott Spillius, ed., *Melanie Klein Today,* vol. 1, 226–248, esp. 231–232.

52. Naess, "Identification as a Source of Deep Ecological Attitudes," 263–264, 268.

53. Naess, "Identification as a Source of Deep Ecological Attitudes," 262.

54. Naess, "Identification as a Source of Deep Ecological Attitudes," 263.

55. Cited in Paul Shepard, "Nature and Madness," in Theodore Roszak, Mary Gomes, and Allen Kanner, eds., *Ecopsychology,* 30.

56. Quoted in Snyder, "Exhortations to Baby Tigers," in *A Place in Space,* 212.

57. Fowles and Horvat, *The Tree.*

58. In Fox, "Transpersonal Ecology and the Varieties of Identification," in Drengson and Inoue, eds., *The Deep Ecology Movement,* 143.

59. Naess, "Identification as a Source of Deep Ecological Attitudes," 262.

60. Shepard, *The Tender Carnivore and the Sacred Game,* 168.

61. Bion, *Cogitations,* 248.

62. Snyder, "Survival and Sacrament," in *The Practice of the Wild,* 184.

63. See D. W. Winnicott, "The Development of the Capacity for Concern," and "Psycho-Analysis and the Sense of Guilt," in *The Maturational Processes and the Facilitating Environment,* 73–82, 15–28. See also Winnicott, "On the Use of an Object and Relating through Identifications," in *Psycho-Analytic Explorations,* 218–222.

64. Winnicott, "The Development of the Capacity for Concern," in *The Maturational Processes and the Facilitating Environment,* 80–81.

65. Shepard, *The Tender Carnivore and the Sacred Game,* 168.

66. Shepard, *The Tender Carnivore and the Sacred Game,* 164–165.

67. On the notion of the transitional object, see for instance Winnicott, "Transitional Objects and Transitional Phenomena," in *Through Paediatrics to Psycho-Analysis,* 229–242.

68. Shepard, *The Tender Carnivore and the Sacred Game,* 166.

69. Shepard, *The Tender Carnivore and the Sacred Game,* 166.

70. This is the general argument of Neville Symington's *Narcissism: A New Theory.*

71. Shepard, *The Tender Carnivore and the Sacred Game,* 164–165.

72. Shepard, *The Tender Carnivore and the Sacred Game,* 169.

3 Tangling at the Roots of Being: Perception as Field and Reciprocity

What I am trying to translate to you is more mysterious; it is entwined in the very roots of being, in the impalpable source of sensations.
—Joachim Gasquet, *Cézanne*

If our relationship to the natural world is one of kinship, reciprocity is what binds kindred beings to one another. What is reciprocity? It is far more than a mere exchange between beings who are essentially external to one another. Let us rather think of reciprocity as mutual recognition and influence that involve not only individual separateness but also a relating based on an embodiment in which psyche and senses, internal and exterior worlds, are engaged. For the philosopher Hegel, it is mutual recognition that allows a person to conceive of himself as a subject. For Donald Meltzer, "aesthetic reciprocity" is the foundation of the baby's development. The baby's self is formed through passionate interchange with the mother's body and psyche, based on the baby's perception of the mother as an aesthetic object—but this is possible only if the mother perceives the baby that way as well.[1] Reciprocity—the mutuality of intimate reciprocal relationship.

In what sense can we understand our relationship to the natural world as one of reciprocity? What could this mean in terms of our relationships with members of other species, with other life forms, with the natural world as a whole? The difficulty of envisioning a reciprocal relationship with the natural world has been a stumbling block in our thinking. To find a way to imagine that relationship as reciprocal is to make it possible to acknowledge our kinship with the natural world again.

Reciprocity as Field and Emergence

> *You know that over there is milkweed from which come glue and string, over the hill beyond that is where the antelopes water . . .*
> —Gary Snyder, "The Real Work"

Gary Snyder notes that Americans go places by going from one point to another, and that in the Neolithic village, space was already becoming linear—you traveled a line from one village to another. But for the Paleolithic hunter, and the hunter who survives as the prototype of the person in relation to the natural environment, space is not a line but a field, and experience takes the form of what Snyder calls a "field sensing of the world": "In a society in which everything comes from the field . . . the landscape with all its wrinkles and dimensions is memorized. You know that over there is milkweed from which come glue and string, over the hill beyond that is where the antelopes water. . . . That's a field sensing of the world."[2]

Snyder's comment points to a crucial aspect of reciprocity: it is a field phenomenon. Reciprocity is not a matter of an external exchange in which one completely separate object encounters and acts on another. Rather, the reciprocity arises within a shared field in which both—indeed, all—parties are participating. In a field, things emerge into our awareness and become focal points. The animal surges up for us. The sound enters our hearing. The breeze blows to and beyond us. Perception—the use of our senses—has this emergent quality. The crucial understanding of perception as experiences emerging within a shared field requires that we see the field as something that palpably contains us, something we are in and part of.

At the same time, it is the coming together of the parties that creates this field. The passionate interchange between mother and baby, Meltzer says, makes a containing space in which their love making can reverberate and escalate.[3] Thus we think too narrowly by far if we imagine reciprocity with the natural world as an "interspecies communication" in which one being expresses itself in its way and then the other expresses itself in another way. In fact the intense mutuality of reciprocity in the containing space within which it occurs inspires awe and amazement. We see this in our response to the mother-baby couple and in the feeling of awe at an encounter in the natural world. Reciprocity is closer to a love relationship than a commodity exchange.

Ultimately, the earth as a whole is the containing field in which our experiences of individual other beings surge up. As David Abram points out, we are not *on* the earth; we are *within* it.[4] If we recognize that the earth is not a flat surface beneath our feet but the whole of the living planet with its creatures and its atmosphere, we can see that the earth encompasses us. Being within the earth in this way, our experience has a quality of depth. Reciprocity within the field has a quality of profound encounter, where mere exchange would be superficial, a mere contact of surfaces. Within the containing space, there is room for resonance and reverberation.

Reciprocity in this sense is intimacy. What emerges into our field awareness emerges into intimacy with us. It is not only the mother, or the infant, or the hunted animal that emerges suddenly in this way. It may be any being that we suddenly become able to perceive. Snyder tells the story of suddenly noticing a tree he has passed many times before: "After twenty years of walking right past it on my way to chores in the meadow, I actually paid attention to a certain gnarly canyon live oak one day. Or maybe it was ready to show itself to me. I felt its oldness, suchness, inwardness, oakness as if it were my own. Such intimacy makes you totally at home in life and in yourself."[5]

The ultimate result of this intimate reciprocity is the growth of consciousness, of human mental structure. Such growth means that one becomes, as Snyder says, more at home both in life and in oneself: the inside and the outside mirror one another, each becoming increasingly complex and ordered as they are increasingly discovered. For if the field in which reciprocity arises can be thought of as a containing space, that space is both an interior and an exterior one. The containing space of intimacy thus comes to be a mental space that can contain experience. One sees this in the prototypical experience of human intimacy—the mother-infant relationship. The aesthetic reciprocity in which mother and baby perceive one another as infinitely rich and full of meaning in both body and mind creates a space in which, as Meltzer says, their love making reverberates as each experience is assimilated and leads on to the next in an ongoing process of growth and development. If this is so, then relating to the natural world is not a diversion from human concerns, not a sacrifice of human growth. Rather, as Laurens van der Post and Basho, among others, have told us, it is a journey to the interior at the same time.

Paul Shepard tells us that he thinks the caves in which the early hunters made their paintings represent the womb of the earth.[6] Their counterpart in the human being, however, he says, is not the womb so much as the skull—the same shape, but now seen from the outside. The memory images within the skull are then the interior counterpart of the paintings on the walls of those caves. Interior and exterior are not only correlates of one another, they are the reciprocals of one another. This is nowhere more beautifully elaborated than in Maurice Merleau-Ponty's discussion of Cézanne's landscape painting.

Painting as Perceptual Reciprocity: Cézanne

We see the world through the eye and not with it.
—William Blake

Nature is on the inside.
—Paul Cézanne

Cézanne's relationship with his subject matter in the natural world shows how very deep the process of reciprocal perception can reach—to the very roots of embodied being, interior and exterior. Aesthetic experience, as in painting, can be an intimate experience of the nonhuman, and reflection on it helps to illuminate our relation to the natural world.

We think of painting as a visual art, but Cézanne's painting, as he understood it and as the phenomenological philosopher Maurice Merleau-Ponty interprets it, shows us that what seems a matter of vision alone is really a matter of reciprocity between one's whole being and the world outside. It is a matter of embodied reciprocal relationship. This is true for our relation to the natural world in general, as we saw. In the field experience the senses are not really separated. We may speak of the voice of nature, or of the hunter's field of vision in which the animal surges up, but the animal surges up as a whole, as a presence in the field. Indeed, the splitting of the separate senses may be a symptom of being lost, as when the autistic children Meltzer studied split their vision and their hearing—"dismantling," as he said, their capacity for experience.[7] Taking Cézanne as the prototype of vision, and elaborating on the statements Cézanne made about his art to his admirers, Merleau-Ponty gives us a phenomenology of this embodied reciprocal relationship.[8]

Cézanne would stand in front of the same scene for hours and return day after day. His emphasis on slow, patient, deep looking is striking, both in his approach to painting and his approach to the natural world. What was he trying to achieve in this way? This was not some act of mere seeing. Rather, Cézanne was endeavoring to let the landscape in its specificity enter into him and be experienced by him, then emerge in the form of a painting. "The landscape thinks itself in me," he said, "and I am its consciousness."[9]

Cézanne spoke of the result he was aiming for as a *"réalisation"*—a realization, or making real, an actualization. Cézanne's relationship to Impressionism on the one hand and to the Old Masters he studied in the Louvre on the other illuminates this idea. From Impressionism Cézanne learned to paint from nature. He rejected this approach, however, because Impressionism restricts itself to vision alone—at least in theory—whereas Cézanne's work is founded on a partnership between nature and the human mind as a whole. To be the space in which the landscape can think itself is quite different from being an eye that looks at nature; to make a painting that is the landscape's consciousness is quite different from recording the way nature appears to the eye.

Cézanne's word *réalisation* suggests that his painting corresponds in some way to the reality of the landscape. While the Impressionists "created pictures," he said, he was "attempting a piece of nature." And while Cézanne admired the Old Masters, he understood himself to be doing something different from them. He was aiming at reality, whereas they "replaced reality by imagination and by the abstraction which accompanies it."[10] He was not, however, attempting the mimetic copying of external reality that we would call "realism." A painting of Cézanne's looks quite different from a photograph of the same landscape. It has a solid reality of its own that is absent from the photograph.[11] Looking deeply into his paintings, we have the sense of a space that reaches out to take us into itself and then moves us back out. (Hemingway, who went to look at Cézanne's paintings every day during his years in Paris, tried to make this movement of embracing space a principle of his writing.[12]) But this movement is not separable from the sense of spatial solidity that the paintings achieve—a solidity that is very difficult to describe. It gives the impression of being the coherence of a quasi-viscous yet nonmaterial substance. Just as Snyder talks about suddenly noticing the oak tree and experiencing its "oakness" in himself, so the link between the viewer's consciousness and stuff in the painting is not separable from the "reality" of the nature equivalent in the painting. Something has been captured, or connected with, internally and transformed into a painting.

For Merleau-Ponty, when Cézanne looks and looks and lets the mountain look back and into him, he is trying to understand, through direct experience, the nature of vision (again, remembering that vision is not mere seeing but an activity of the embodied senses): "It is the mountain that he interrogates with his gaze. What exactly does he ask of it? To unveil the means, visible and not otherwise, by which it makes itself a mountain before our eyes."[13] This process of vision is then to be represented in his painting. This for Merleau-Ponty is the essence of Cézanne's enterprise.

Merleau-Ponty takes this sense of a reciprocal movement between interior and exterior as the basis for explaining his understanding of perception, and this is why Cézanne's work is so important to him. Vision, indeed all perception, he says, is not a matter of a subject perceiving an object but rather something more complex—a chiasmus, a movement from outside to inside and back again, on both sides, in both directions. "Vision happens among, or is caught in, things," he says, "in that place where something visible undertakes to see, becomes visible for itself by virtue of the sight of things; in that place where there persists . . . . the undividedness of the sensing and the sensed."[14] This is the nature not only of perception but of bodily—fleshly—existence as such. "The painter takes his body with him," Paul Valéry had said. He meant, Merleau-Ponty explains, that we cannot imagine a painting without a body— without hand and eye—or without movement. But the very body that moves and that the painter uses to descend to the depths of being in trying to grasp how sensation and perceptual experience arise—that body is made of the same stuff as the world. It is flesh as the world is flesh. Not only that, it is perceptible—"*sensible,*" sense-able, Merleau-Ponty says—as the world is. It can be seen, just as the embodied eye can see the world: "Visible and mobile, my body is a thing among things; it is caught in the fabric of the world, and its cohesion is that of a thing. But because it moves itself and sees, it holds things in a circle around itself. Things are part of its full definition; the world is made of the same stuff as the body."[15]

Painting brings to our attention this complex reciprocal interplay of flesh and vision, the visible and the seen, the interior and the exterior. Vision happens when the exterior things resonate with their internal trace (or "carnal formula," in Merleau-Ponty's phrase) in the body. "Since things and my body are made of the same stuff, vision must somehow take place in them; their manifest visibility must be repeated in the body by a secret visibility," writes Merleau-Ponty.[16] It is as though the painter who tangles at the roots of being perceives with an internal vision that is the reciprocal of what appears out side: "Quality, light, color, depth, which are there before us," he continues, "are there only because they awaken an echo in our body and because the body welcomes them." This reciprocity of interior and exterior vision is what Cézanne is referring to, I believe, when he says, "Nature is on the inside." The painting that we see has emerged as an outward version of those "inward tapestries," as Merleau-Ponty calls them,[17] those internal traces of outward visibility. It allows us to see them, and in doing so it teaches us what seeing is. It shows us how in the very act of seeing and being seen we are inextricably entangled with the outer world.

In this process, there is a clear sense of initiative on the part of the visible in external nature. Merleau-Ponty notes how frequently painters say that things are looking at them. "In a forest," says Paul Klee, "I have felt many times over that it was not I who looked at the forest. Some days I felt that the trees were looking at me."[18] If Cézanne stood for hours in front of the Montagne Sainte-Victoire and drew and painted it many times, "it is the mountain itself which from out there makes itself seen by the painter," Merleau-Ponty explains.[19] We imagine that painters look at their motif, not the reverse. Here we see that the movement goes in both directions in this tangling at the roots of being.

The Hunter and the Open Mind

> *The hunter knows that he does not know what is going to happen, and this is one of the greatest attractions of his occupation.*
> —Ortega y Gasset, *Meditations on Hunting*

Looking is indeed one of our main activities in relation to the natural world, and Merleau-Ponty's discussion of Cézanne helps us reflect on the nature of looking. Looking is the hunter's central activity as well. The hunter in the landscape watching for the game he seeks is the analogue of Cézanne the painter looking to tangle with his motif. Like Gary Snyder, we will take the hunter as the ideal type of the person who engages in a relationship of perceptual reciprocity with the natural world. (But when I say "the hunter" I do not mean the trophy hunter, alienated product of an alienated culture, but rather the hunter in a hunting-gathering society, bearing in mind at the same time that there are wide variations among such societies and such hunters.)

Remember what Valéry told us: "The painter takes his body with him." We are not concerned here with vision only but rather with full embodiment.

Looking is not vision alone, nor is perception separable from engagement and relationship. Full embodiment in the context of the natural world, as we see in the case of the hunter, consists of an openness to what emerges. What characterizes the hunter's way of looking is alertness. "Imitating the perpetual alertness of the wild animal, " says José Ortega y Gasset, philosopher of the hunt, the hunter "sees everything."[20] He sees everything because he does not have preconceptions. He is ready for the game to surge up into view unexpectedly: The hunter "does not believe that he knows where the critical moment is going to occur. He does not look tranquilly in one determined direction, sure beforehand that the game will pass in front of him. The hunter knows that he does not know what is going to happen, and this is one of the greatest attractions of his occupation."[21] The hunter's gaze, in other words, places him *within* a field in which contact with the prey can be made at any time. Withinness is a prerequisite for contact; no contact is possible if one is not *within* the same context—if the world and the body are not made of the same stuff.

Ortega y Gasset talks about hunting as a way of reconnecting with our animal state. He calls this "a vacation from being human," but we might think of it as a way of remembering our continuity with the natural world. "Man cannot reenter Nature," he says, "without temporarily rehabilitating that part of himself which is still an animal. And this, in turn, can be achieved only by placing himself in relation to another animal. But there is no animal, pure animal, other than the wild one, and the relationship with him is the hunt."[22]

The hunter, then, makes contact with the true animal, the wild animal, and the point of contact is the hunt. The hunter's alert gaze allows this contact. It links the hunter with the animal and the place, all within the same shared field. The field that the hunter shares with the prey Ortega y Gasset calls "countryside," as opposed to landscape. "Only the wild animal is properly in the countryside, not just on top of it, simply having it in view," he says.[23] Thus "only the hunting ground is true countryside," and "only by hunting can man be in the country; . . . within a countryside which, moreover, is authentically countryside."[24]

This sense of being within the countryside depends on the reversal of perspective that occurs when the hunter starts to see as the animal does, for hunting is an imitation of the animal. The hunter says Ortega y Gasset, "will instinctively shrink from being seen; he will avoid all noise while traveling; he will perceive all his surroundings from the point of view of the animal, with the animals' peculiar attention to detail."[25] This is what he means by being *within* the countryside. This is no mere external imitation, however, but rather a linking of the exterior and the interior as the hunter identifies with the animal. The imitation creates true reciprocity: the pursuer cannot pursue if he does not integrate his vision with that of the pursued. The hunt links them in the medium of alertness.

This withinness is the basis for the kind of field perception we have been talking about. At the same time, being together in the field guarantees true individuality to both hunter and prey, and to place as well, for there is no "nature in general" but only this specific field that the hunter and the animal constitute between them, which is this specific countryside and every place and every moment in it. Just as the hunter's alertness is in the service of the moment when he will act to take the prey, so the individual elements of the countryside act in concert in the field of the hunt. Ortega y Gasset evokes the richness of this field:

> Only when we see [the countryside] through the drama that unfolds in the hunt can we absorb its particular richness. Articulated in that action, . . . wind, light, temperature, ground contour, minerals, vegetation all play a part; they are not simply there, as they are for the tourist or the botanist; but rather they function, they act. And they do not function as they do in agriculture, in the unilateral, exclusive, and abstract sense of their utility for the harvest, but rather each intervenes in the drama of the hunt from within itself, with its concrete and full being.[26]

Thinking about the painter's and the hunter's way of looking alerts us to what is going on in more alienated ways of looking. Ortega y Gasset contrasts the hunter's open field of visual awareness—his alertness—with the gaze of the tourist. While the hunter is within countryside, the tourist sees landscape: "The tourist sees broadly the great spaces, but his gaze glides, it seizes nothing, it does not perceive the role of each ingredient in the dynamic architecture of the countryside."[27] The hunter, as opposed to the tourist or the narrowly focused farmer who remains outside the countryside, has a quest that compels his alertness. The painter is like a hunter; he wishes to capture, on the inside, his motif. Cézanne interrogates the mountain, and he needs to look until that tangling has made a connection at the roots of being. Tourists, however, have no quest (if they do, they are no longer tourists but travelers). Hence their looking is without alertness; there is nothing that can surge up to be captured. They have no orientation from which the interrelations of the surroundings can be perceived, just as they have no point of contact that would situate them within the countryside.

How foolish, then, to think—or to try to convince ourselves—that watching a video about a wild animal is somehow similar to an encounter with that animal in the wild. The tourist and the videowatcher share a mere looking that is external and that allows for no point of contact with what is seen. Unlike Valéry's painter, they do not take their bodies with them. The body of the videowatcher and the body of the animal filmed do not inhabit the same context; in fact, few of the videowatcher's senses are involved. It is possible to imagine that there is some link with nature in videowatching only because videowatching provides a misleading sense of presence that obscures the actual sensory deprivation involved. In the wild, the animal emerges—surges up, as Ortega y Gasset would say— for only a moment, to provide a fleeting glimpse. The hunter must be alert so that he can be surprised and act out of the immediacy of the moment. The animal in the video, however, is unnaturally present and unnaturally under our control. Its apparent availability is the sign of its actual unavailability for real contact. As Snyder notes, "Wildlife is often simply a call, a cough in the dark, a shadow in the shrubs. You can watch a cougar on a wildlife video for hours, but the real cougar shows herself but once or twice in a lifetime."[28]

The hunter in a hunting-gathering society, considered as the ideal type of the person who "tangles at the roots of being," hunts not only to kill but also to eat what he has killed. Ortega y Gasset neglects this crucial aspect of the hunt. For Paul Shepard, however, who has also meditated deeply on hunting, the hunt is real only if the prey is killed and eaten. "Among hunters the fallen prey is universally cherished and honored both for itself as a thing of beauty and in the solemnity and meaning of its death," he reminds us. "Ecologically, death leads to life," he continues, "not in a hazy and obscure way, but in the eating of the prey. Men must eat to live. Men must also eat [the prey] because it is wrong to waste the life of the prey."[29] Eating the prey provides the link between death and life and makes the hunt a meditation on life and death. To eat what we have killed is to take into our bodies something we confirm. The hunter becomes still more deeply engaged in the bodily interchange of life and death. The hunt becomes, again in Shepard's words, "the symbol par excellence of life, which is used—and uses him—in the process."[30]

Through the meditation that is the hunt, the life of the living world outside and the life of the living psyche inside interpenetrate. We see this in the

Papago hunter's quest for song: "A man who desires song did not put his mind on words and tunes. He put it on pleasing the supernaturals. He must be a good hunter or a good warrior. Perhaps they would like his ways. And one day in natural sleep he would hear singing. He hears a song and he knows it is the hawk singing to him of the great white birds that fly in from the ocean."[31] In the hunt, the hunter and the prey use and are used in turn by life as this process of interpenetration. Just as the hunter uses the prey to live, and Cézanne uses the mountain to think about the nature of vision, life uses the hunter's psyche to let the ocean birds sing, precisely as life uses Cézanne to bring the Montagne Sainte-Victoire to consciousness in his painting.

The Hunter and the Terrain of the Interior

> *Hunting is one of the hardest things even to think about.*
> —Edward Abbey

> *. . . dapple-dawn-drawn Falcon, in his riding*
> *Of the rolling underneath him steady air . . .*
> —Gerard Manley Hopkins, "The Windhover"

Ortega y Gasset's book is called *Meditations on Hunting*—it is a work of philosophy. Plato and Socrates must themselves have been hunters, Ortega y Gasset tells us; they referred to the philosopher as a hunter. In fact the philosopher needs to have the same kind of open alertness to what emerges as the hunter of animals: "Like the hunter in the absolute outside of the countryside, the philosopher is the alert man in the absolute inside of ideas, which are also an unconquerable and dangerous jungle." Not only do ideas inhabit the wild terrain of the inside as the animal inhabits the wild terrain of the outside; the kind of attitude the philosopher requires is precisely analogous to that of the hunter. "In fact," Ortega y Gasset says, "The only man who truly thinks is the one who, when faced with a problem, instead of looking only straight ahead, toward what habit, tradition, the commonplace, and mental inertia would make one assume, keeps himself alert, ready to accept the fact that the solution might spring from the least foreseeable spot on the great rotundity of the horizon."[32]

The alertness of the open mind, and the interpenetration of interior and exterior that it implies, links the hunter with the painter and the philosopher—and with the psychoanalyst as well. The psychoanalyst is par excellence the person who takes the interior of the psyche in its vastness and its developmental potential as his or her terrain, or hunting ground. This is most articulately evident in the thinking of Wilfred Bion, who extends Freud's notions of an "evenly hovering attention" and of deliberately blinding oneself in order to be able to see the one spark of light in the darkness. Bion recommends that the analyst deliberately refrain from memory, desire, and understanding—from the memory of what the patient has said earlier and the "facts" of the patient's life, from the desire to "cure" the patient, and even from the activity of making sense of what he or she hears and sees.[33] This does not mean that the analyst must become stupid; the goal is to remain in the open field of the present, rather than being drawn away into the past, through memory, or the future, through desire, or into preconceived ideas, through an effort to understand. The analyst needs to remain alert in a state of not-knowing, open to the further emergence and evolution of what Bion calls "O," the ultimate (and unknowable) reality that, unfolding, realizes itself in the psyches of patient and analyst.

It is especially in this utter rejection of preconceptions and premature closure that Bion's attitude resembles that of the hunter. As Ortega y Gasset says, the hunter never knows what will happen next, and thus "he needs to prepare an attention of a different and superior style—an attention which does not consist in riveting itself on the presumed but consists precisely in not presuming anything and in avoiding inattentiveness."[34] In this state of not-knowing, there is a tremendous uncertainty and risk; it is as though all the analyst has to go by is what the Taoist Chuang Tzu called the "torch of chaos and doubt."[35] As Aristotle said in the *Nicomachean Ethics,* "it would seem that the experience of particular things is a sort of courage."[36]

The analyst waits in this state of "unknowing" for "a pattern to obtrude," as Bion puts it, following Charcot. At that point the prey has emerged, and the hunter acts; something has crystallized in the analyst's mind out of the field of unknowing. Like the arctic hunters Hugh Brody studied, the analyst then becomes like the falcon: "Above all they are still and receptive, prepared for whatever insight or realization may come to them, and ready for whatever stimulus to action might arise. This state of attentive waiting is perhaps as close as people can come to the falcon's suspended flight, when the bird, seemingly motionless, is ready to plummet in decisive action."[37]

For the analyst as for the hunter and the painter, the uncertainty and the danger are accompanied by a faith in life that is the foundation for this openness, a faith that life will provide, even as it uses the person for the ends of life itself. The mind open to the emergence and evolution of psychic reality both cherishes and confirms that reality and is nourished by "eating" it. Emerging reality is the food of the mind.

Coevolution: The Turbulence of Reciprocal Shaping

The rhythm of streetcars and carpet-beating rocked me in my sleep. It was the mold in which my dreams took shape.
—Walter Benjamin, *A Berlin Childhood ca. 1900*

The Winds that enter a human being leave their trace, according to the Navajo, in the vortices or swirling patterns to be seen on our fingertips and the tips of our toes, and in the spiraling pattern made by the hairs as they emerge from our heads.
—David Abram, *The Spell of the Sensuous*

In his book *The Spell of the Sensuous*, David Abram elaborates on this vision of an interpenetrating perceptual reciprocity. He evokes a world in which every aspect of our being is nourished and molded by our sensuous engagements with other creatures, a world in which these relationships have shaped us and made us what we are. "Humans are tuned for relationship," Abram writes, suggesting that "we are human only in contact, and conviviality, with what is not human. . . . The eyes, the skin, the tongue, ears, and nostrils—all are gates where our body receives the nourishment of otherness. This landscape of shadowed voices, these feathered bodies and antlers and tumbling streams—these breathing shapes are our family, the beings with whom we are engaged, with whom we struggle and suffer and celebrate."[38]

As we have seen, this sensory engagement is not a collection of perceptions in different sensory modalities. As an engagement of living beings, it is multisensory, indeed synesthetic. The sensing body synthesizes the various sensory aspects of the experience of another. In Gasquet's phrase about Cézanne's work, it "joins the wandering hands of nature." "The raven's loud, guttural cry, as it swerves overhead," writes Abram, "is not circumscribed within a strictly audible field—it echoes through the visible, immediately animating the visible landscape with the reckless style or mood proper to that jet black shape. My various senses, diverging as they do from a single, coherent body, coherently converge, as well, in the perceived thing, just as the separate perspectives of my two eyes converge upon the raven and convene there into a single focus."[39]

Through this synthesis, the perceived thing becomes a coherent object that is dynamic and alive, a "center of forces" or a "nexus of experience," in Abram's words—an Other. This is equally true for our interior experience. Wilfred Bion speaks, tongue in cheek, of "common sense" in experiencing the life of the psyche.[40] Just as an external object is perceived only when the various sensory experiences of it are synthesized, so the experience of an interior or psychic "object" becomes coherent only when the various emotional experiences of it are united in our conception of it. How could we fully relate to another person only through sight or sound—or love or hate? To experience only one aspect or only through one experiential modality, is to be disconnected, "dismantled," as Meltzer says of autistic children.

This synthesizing activity is impelled and invited by the experienced being—the raven "gathers my senses together," says Abram—and encouraged by our participation in the life process. To make coherent is to make individual and located. We are able to perceive that particular raven, to experience that particular person. The more we are able to gather experience together in this way, the more deeply we are engaged in the process of life, and the more we become part of the vast web of life. With this understanding we can envision the world as a whole as a "vast interpenetrating webwork of perceptions and sensations, borne by countless other bodies"—so that our sensory perceptions are supported as much by the motion of the owl's wing or the tumbling of a stream down a slope as by our own capacities.[41] It is this interpenetrating of nexuses that is the true nature of "reciprocal" perception—much more fully intertwined and interrelated than any conception of reciprocity as simple exchange would suggest.

Experienced in this way, the nonhuman is not Nature as an abstract concept but rather the "living land," as Abram calls it, a spacious present, a field that we are within, full of motion and engagement. The wind, the "unseen, imperturbable wind" (Abram again) can stand as the prototype of this interpenetrating webwork of nexuses. For every motion of any kind is felt in the air, creating whorls and turbulences that are carried beyond us, to interpenetrate with the motions of others. Our breath, our speech, the cry of the raven, the flapping of a wing—all these set the air in motion and are felt by one another.[42]

This turbulence that is our engagement in the living land does not stop at our bodies but extends into our psyches as well. Our emotional engagements create the same kind of turbulence in our interior experience. (Bion uses Leonardo da Vinci's drawings of hair with its whorls as an image for this turbulence.) Every encounter of one personality with another sets the psyche in motion in complex ways. "When two personalities meet, an emotional storm is created," is Bion's way of putting it.[43] Just as the Papago hunter's experience of hunting initiates a process in his psyche through which he receives song, so our encounter with the unseen psyches of others initiates reverberations in vast webworks of experiences from whence arise actions in ourselves that reach out to affect others.

In Abram's view, evolution is best understood as coevolution—the way coexisting, engaged beings shape each other in an unfolding and unpredictable process. Our participatory engagement has formed us along with all other life forms. It is because we have patterned one another over time that our senses can engage in the reciprocal engagement that is perception: "These other shapes and species have coevolved," Abram writes, "like ourselves, with the rest of the shifting earth; their rhythms and forms are composed of layers upon layers of earlier rhythms, and in engaging them our senses are led into an inexhaustible depth that echoes that of our own flesh."[44] This is the sense in which Cézanne experiences the interior traces of what he sees outside himself. In this sense nature is truly on the inside as well as the outside.

The natural world as a whole provides the stability within which this shifting and shaping can proceed, for nature evolves slowly enough for us to participate in these nexuses and be shaped by these rhythms. Rhythm creates form, and we are able to perceive common form in a natural world that moves at the pace of evolution. On the basis of this foundation we can engage in a perceptual experience that is constantly alive and unfolding, a kind of minievolution. Abram again, evoking as he does so well this sense of unfolding aliveness: "The patterns on the stream's surface as it ripples over the rocks, or on the bark of an elm tree, or in a cluster of weeds, are all composed of repetitive figures that never exactly repeat themselves, of iterated shapes to which our senses may attune themselves even while the gradual drift and metamorphosis of those shapes draws our awareness in unexpected and unpredictable directions."[45]

Things produced by processes less rich than the reciprocal patterning of coevolution are less nourishing to the mind and the senses. The farmer as Ortega y Gasset understands him has a less fully participatory mind because he has narrowed his interest to specific utilitarian functions that the landscape may play in the production of his crops. For the most part, human-made and especially mass-produced objects offer a narrower scope for perceptual engagement by virtue of the processes by which they are made. Such objects are preconstrained, as it were. We do not even have to restrict our awareness to take them in, for they have not been shaped by rhythmic variation within a framework of stability. Rather, they bear the marks of a repetition *without* variation. They are so unvarying and so little differentiated that they lose the vitality of "eachness," uniqueness. Such objects cannot surprise us or participate with us in the openness and creativity of the vast learning experience that is coevolution. Once our bodies have mastered the specific functions for which they were designed, these mass-produced artifacts of civilization have nothing left to teach us.[46]

James Hillman puts the plight of these human-made, mass-produced things more compassionately: "Things are composed of poisonous and flammable substances, stamped out of uniform molds, internally fastened cheaply, quickly with the least care, untouched by the human hand. They cannot weather or age. Their existence is hurried by the push of obsolescence as one generation succeeds the next within a few months. Sold by barkers on the slave-blocks of the market, . . . their suffering is written on their faces."[47] His compassion reminds us that human-made objects too can be vital and can engage in perceptual reciprocity with us. Cézanne's paintings have certainly shown us that this is the case.

Life Lessons from Other Species

> *The mountains, the creatures, the entire nonhuman world is struggling to make contact with us.*
> —David Abram, "Notes from Thami Valley, Nepal"

In a lesson from another species that we may read as quite literal or as symbolic, an elderly Papago man, as we heard earlier, brought a mockingbird home to feed to his mute son, in order to impart to the boy the quality of loquaciousness.[48] In a more contemporary incident of learning from experience with another species, Freeman House, whose book *Totem Salmon* bears the subtitle *Life Lessons from Another Species,* describes a profound encounter with the salmon during a youthful summer on a fishing boat. Fish brought up with nets were lying on the deck and being swept into the hold. There was no time to attend respectfully to their individual dyings. House has no memory of how he came to do what he did next, he tells us, but "I found myself alone in a dark corner of the fish hold squatting with a ten- or twelve-pound sockeye salmon still alive across my knees. With a knife I opened up its chest cavity just enough to find the heart and tear it out with two fingers and a thumb. It came with a ripping and squirting sound. I popped the heart into my mouth and bit down once, hard, through the gristly thing."[49]

House spontaneously enacts something that is built into hunting rituals—he takes the heart of the hunted animal, along with its qualities, into himself. He does this outside the context of tradition, and inside the context of an economic fishing enterprise. This contact with the salmon's heart is an experience of terror for him, terror evoked by the radical otherness of the nonhuman Other: "One bite brought a flavor like all of Icy Straights and enough saliva to float ten salmon hearts, enough to swallow it whole. As I swallowed, all my floating terrors gained a name and swept through me with the intensity of a hurricane. Fear of fish flesh and cold blood, fear of slime on the flesh of fish and its absence on the skin of snakes, fear of the strangeness of other species, fear of a world barren of human thought, fear of death: my own and all of it."[50]

But along with the terror, the experience brings a greater clarity and awareness. It shifts, though as yet obscurely, his sense of who and where he is: "I resumed the drill of preparing for the next set with an unusual clarity of vision and emptiness of mind. The shards of light reflecting from the surrounding sea, the crazy screams of the gulls and terns, the fading colors of the dying fish, all took on the aspect of a single thought that may or may not have been my own."[51] A life lesson from the salmon: for House, writing years later, the experience evokes David Abram's words: "The whole Earth is rumbling and straining to let us remember that we are of it, . . . that the Earth is our real body and that it is alive."[52]

This life lesson began as a single dramatic experience, but it continued to reverberate. Reading *Totem Salmon,* one feels that this incident was the nucleus from which emerged the long life story of House's work on salmon restoration chronicled there. Real contact with another species changes our awareness, dramatically. We are no longer so separate, and our being is fundamentally altered. The unfolding continues from this shift. That is another part of the lesson of the salmon's heart.

How Animals Made Us Human

Our reflections on how the mind works must . . . carry us beyond ourselves,
pursuing the nature of thought as the thought of nature.
—Paul Shepard, *Thinking Animals*

It is Paul Shepard who has given the most explicit and detailed attention to the
ways human interaction with the natural world, and particularly with animals,
is a learning experience. His book *The Others: How Animals Made Us Human* is
a virtual catalog of the cultural and psychic uses we make of animals. What
emerges from his examination is a picture of the way human development
makes use of the natural world at every stage, from infancy to full maturity.
If gifts create bonds, as Lewis Hyde tells us in his book *The Gift,* this use of
the gifts of nature becomes the basis of our developing bond with the natural
world. "Only in the success of this extraordinary calendar," Shepard writes,
"does the adult come to love the world as the ground of his being."[53]

Shepard draws on his understanding of hunting-gathering societies to outline
the ideal-typical calendar of human development. The natural world is used for
both emotional and cognitive development. Young children in their imitative
play draw on the various qualities of the animal world around them, qualities
they will also find incorporated in the dances of the adults. Through what they
see around them, they are able to grasp the individual qualities of feeling and
action that they might find in themselves at a particular time—a "personal inner zoology." "Play is an imitation," Shepard explains, "starting with simple
fleeing and catching, going on to mimic joyfully the important animals, being
them for a moment and then not being them, feeling as this one must feel and
then that one, all tried on the self."[54] This is using animals, through play, for a
knowledge of the internal world of the self. But for children, animals are also
concrete physical beings. They are used in all the variety of their physical form
for the development of thinking. Their similarities and differences provide the
material for the categorical thinking that Piaget calls *concrete operations.*

Adolescence, in contrast, coincides with the development of what Piaget calls
formal operations. The internal world opens up, and abstract, philosophical,
metaphysical thought becomes possible. The initiation ceremonies of adolescence put the natural world to a new use: nature is contemplated as "a poem,
numinous and analogical, of human society." The adolescent does not leave his

childhood interests in the natural world behind, Shepard tells us, so much as graduate into its significance as the metaphorical correlation of interior and exterior worlds and their emergent quality become apparent. Adolescence ushers in, in his words, "a lifelong study, a reciprocity with the natural world in which its depths are as endless as his own creative thought." Maturity represents a continued deepening of this reciprocity rather than an alienation. It "celebrates a central analogy of self and world in ever-widening spheres of meaning and participation," Shepard writes, "not an ever-growing domination over nature, escape into abstractions, or existential funk."[55]

Contrasted with the developmental design of hunting-gathering societies as Shepard portrays them, the truncation and distortion of others become apparent. It is as though with each phase of Western European culture we limit ourselves to earlier and earlier stages of development: this is the argument of Shepard's illuminating and frightening book *Nature and Madness*. As we fail to provide the means for establishing each phase of the evolving relationship to the natural world at each stage of the person's development, we increasingly fail to reach true adulthood. Lacking full adolescent rituals, we rarely make it past childhood into maturity. There are then few genuine elders to help those in the next generation, who in turn rarely become adults who can guide children through the developmental stages.[56] Hence we easily remain caught up in the fears and fantasies that belong to what Shepard calls "the immense concerns of immaturity: separation, otherness, and limitation"—childhood fears we lack the means to grow out of. "The West is a vast testimony to childhood botched to serve its own purposes," he tells us, "where history, masquerading as myth, authorizes men of action to alter the world to match their regressive moods of omnipotence and insecurity."[57]

From within our limited and limiting cultural perspective, we imagine ourselves nostalgic for the richness of childhood. But Shepard's vision informs us that it is not childhood we have lost, but maturity. In childhood the natural world is grasped in its concreteness, but not yet in its full participatory resonance with the human psyche. With so little conception of that fuller resonance, we do not long for it. Similarly, in the absence of a cultural framework to make sense of them, Westerners' isolated experiences of real contact with the natural world are hard to understand and assimilate. Taking place outside a cultural context, House's experience with the salmon heart required a lifetime of work before it could be understood for what it was, a life-defining lesson from another species.

Learning from Experience: The Baby and the Artist

Every object properly perceived opens up a new organ of perception in us.
—Goethe

The artist who paints a little street in Delft can see and communicate a reality to the observer who then sees something that is quite different from any brick wall or little house that he has ever known or seen in his life.
—Wilfred Bion, *Cogitations*

If we reflect on how much learning goes on within the relationship between mother and baby, it should not be so surprising that the love of nature means learning from nature. The kind of interweaving participatory reciprocal perception we have been talking about is a dense learning environment.

Let us look more closely at this notion of learning in relationship. If coevolution means shaping and being shaped by the rhythms in which we participate, it is very much analogous to the way the new baby's mind is formed in interaction with the mother's mind and body—with the rhythms of interaction that develop and reverberate between them. In Bion's model the mother's more developed and containing mind takes in and makes meaning of the baby's relatively unprocessed emotions. The baby can then assimilate the emotions, now given meaning and context, and on the basis of the internal form thus created go on to the next step in growth and development. This is an ongoing process, but it unfolds from what is created at each point. This radical newness that emerges from the engagement of minds is what Bion calls "learning from experience."

Goethe tells us that an object properly perceived shapes our capacity of perception in new ways. The baby's mind and capacities "grow," so to speak, through this reciprocal engagement with another mind. Similarly, Cézanne, in Merleau-Ponty's formulation, "learns to see by seeing," through the process of being open to what emerges in his encounter with what is outside himself.[58] He does not impose "his vision" on his subject matter; rather, just as the landscape thinks itself in him, so the meaning of whatever he sees emerges within his vision. Again, as Blake said, we see through the eye and not with it. What is grasped emerges within this deep contact, as though from the other itself: "The meaning Cézanne gave to objects and faces in his paintings presented itself to him in the world as it appeared to him. Cézanne simply released this meaning: it was the objects and the faces themselves as he saw them which demanded to be painted, and Cézanne simply expressed what they wanted to say."[59] The meaning of the thing seen becomes Cézanne's own vision.

We would be completely wrong, then, to imagine that the work, the painting, exists first in the form of a conception, an idea. The meaning contained within the painting is an emergent meaning. Preconceived meaning cannot *emerge* in this same way. The meaning of what the artist is going to say does not exist anywhere. As Merleau-Ponty insists, "There is nothing but a vague fever before the act of artistic expression, and only the work itself, completed and understood, is proof that there is something rather than nothing to be said."[60]

The meaning that the viewer of the painting experiences is similarly a matter of reciprocal perception. It emerges from a deep experiential contact between viewer and painting. The work teaches the viewer how to see it, as Merleau-Ponty says, and the viewer looks at the painting with newly emerging perceptual capacities. It is only in this way that one can learn from a painting, or another object or being: "It is not enough for a painter like Cézanne, an artist, or a philosopher, to create and express an idea; they must also awaken the experience which will make their idea take root in the consciousness of others. A successful work has the strange power to teach its own lesson."[61]

Part of the success of the work, in other words, consists precisely in its ability to do this kind of teaching. Just as the earth calls out to us to show us that it is alive and that we are part of it, so the work of art calls out to our minds that we are all part of one field, united for the moment in the emergence of this new capacity. When the painter's image comes to life in the viewers, Merleau-Ponty says, "the work of art will have united these separate lives; it will no longer exist in only one of them like a stubborn dream or a persistent delirium, nor will it exist only in space as a colored piece of canvas. It will dwell undivided in several minds, with a claim on every possible mind like a perennial acquisition."[62]

This emergence of something new entails at the same time a profound change in the individual. Bion calls it a "catastrophic" change, emphasizing how profoundly the individual has been disrupted. The new way of seeing that is revealed in painting is a psychic as well as a perceptual way of seeing. New capacities restructure the psyche, and the newness of this way of seeing represents psychic growth and development, psychic change, a change in the level and kind of relatedness we experience. Think of how dramatically the baby's mind is changed by the emergence of new concepts and capacities, how much of the old way of being is lost when the self is restructured under the impact of learning from experience. Just as Bion recommended that the therapist suspend memory, desire, and even understanding in order to let the evolving reality emerge into awareness, so Cézanne had to lose himself, descend to "the roots of being," and then return, new.

This disintegration of previous structures is what enables a new synthesis to occur, just as our own senses synthesize the various aspects of the perceived object. In order to "join the wandering hands of nature," as his admirer Gasquet put it, and find "the landscape in its totality and in its absolute fullness,"[63] Cézanne had to radically suspend his previous understanding of what he was looking at. According to Mme. Cézanne, he would start by investigating the geology of the landscape and then he would halt and look at everything with widened eyes, "germinating with the countryside" until he was able "to recapture the structure of the landscape as an emerging organism."[64]

Cézanne's painting teaches us all a new way to see, and in the process it teaches us that seeing is a matter of reciprocal perception. The landscape is not simply something out there to be seen like an object but something that both forms itself in our sight and gives us, so to speak, our capacity to see by virtue of its being seen by us. Tangling at the roots of being entangles us with the objects of our vision in the evolution that is reciprocal shaping.

Tradition and Emerging Reality

> *A successful work has the strange power to teach its own lesson.*
> —Maurice Merleau-Ponty, "Cézanne's Doubt"

> *There is a secret person undamaged in each of us, aware of the validity of [the] conditions [for becoming fully human], sensitive to their right moments in our lives.*
> —Paul Shepard, "Nature and Madness"

"The answer is the misfortune of the question," says Bion, quoting Blanchot.[65] Is this notion of the radically new that shapes us in turn at odds with the notion of a tradition, a cultural context, as in hunting-gathering societies, that helps us to assimilate our encounter with the natural world? Are human societies necessarily at odds with wildness? Or is tradition crucial if we are to be open to the radically new? Perhaps it is simply European-American (or industrialized, or urban) societies whose traditions and culture are antithetical to the newness that can emerge through an encounter with the natural world. Psychoanalysis too, which in its now one hundred years has built up a culture of its own, Bion suggests, may become a set of ideas that plugs up the holes that questions might make, rather than a way of opening ourselves to the new.

The case of Cézanne shows us that that there is a double movement involved in the artist's relation to tradition. "Art is built directly upon previous art, indirectly on the artist's courage about himself and about his surroundings," Adrian Stokes tells us.[66] Artists learn from traditions, and only afterward suspend what they have learned in order to open themselves to what emerges. Merleau-Ponty reminds us that when Cézanne was in Paris he went to the Louvre every day because "one must learn how to paint."[67] He studied the geology underlying the landscape and then suspended his knowledge of it. Similarly, it is only the experience they have acquired that permits psychoanalysts to suspend memory and desire in their courageous attempt to receive the newness that emerges at each moment in their work with patients.

Emergence is an unfolding against the background of what has come before, a background that has stability and a rhythmic pattern. Shepard points out that the stability of the natural world, with its slow pace of change, allows for the stability of culture and tradition in hunting societies[68] (or at least it did so prior to contact with Europeans). The natural world is the background, so to speak, against which rituals unfold and personal relations are played out. This stability makes the emergence of the new and the individual possible, just as coevolution occurs through an underlying rhythm of repetitive patterning in which slight variations result in uniqueness. Tradition sets the expectation that an individual undertaking a vision quest will have dreams and find a spirit ally, but it does not predetermine the content of the dreams or the specific songs given by the animal ally. The stability of culture, grounded in the stability of nature, allows adolescents to look ahead to the vision quest with both trepidation and anticipation—in the knowledge that the elders have gone through it before them and will provide them with the proper preparation. We use, as we are used by, tradition. Emergence remains possible, protected by the cultural setting.

And of course traditions and rituals can fail us. Cézanne had a tradition to use, visible in the works of the Old Masters in the Louvre. But it is not at all clear that those of us who are contemporary Euro-Americans have a tradition that will sustain us in our efforts to return to alert and open participation in the patterning of reciprocal perception. There are only reminiscences of exchange with the natural world in most of the fragmentary rituals and traditions we still enjoy, moving though they may be for some participants (think of graduation ceremonies, for instance, with their mixtures of triumph, gratitude, and cynicism felt in various degrees by the individual participants). The short-lived styles and fashions reflected in our manufactured objects may in fact be the quintessential form of contemporary tradition. The large proportion of ugliness in mass-produced objects, their early obsolescence, and their lack of spontaneous variation reflect the rigidity of the patterns of thought that have shaped them. There is little evidence in them of the turbulent coevolution that arises from a tangling at the roots of being. Unlike Cézanne's paintings, they fail to nourish and teach us.

Bion asserted long ago that our capacity to transmit technical knowledge through mimicry was far in advance of our capacity to communicate emotional knowledge, which cannot be transmitted by imitation. Perhaps it would not be fair to call the unvarying patterns of thought that shape so many of our objects "tradition." But let us ask with Bion how we might transmit an emotional knowledge that would allow us once again to be open to encounters with the natural world. In what would that knowledge be based? And what form would its transmission take?

Our most thoughtful environmental philosophers have struggled with this question. Shepard may have the most encouraging response to it. The desire for such facilitating traditions, he argues, is built into the human genome. We long, in other words, for the kinds of mentoring and ritual activities that would help us reach the next stage of our development, even as we know intuitively that those activities cannot be prefabricated but must themselves emerge from experience. "The problem may be more difficult to understand than to solve," Shepard writes: "An ecologically harmonious sense of self and world is not the outcome of rational choices. It is the inherent possession of everyone; it is latent in the organism, in the interaction of the genome and early experience."[69] The natural world in each of us—here represented by the genome—continues to be available as the potential foundation for a culture that would foster and shelter the emergence of the new.

The Gift and the New

> *If the purse seine boat was a model of the commodity economy, salmon had shown me that it was floating in a sea of natural provision, the boundless generosity of the Earth.*
> —Freeman House, *Totem Salmon*

In his lovely book *The Gift: Imagination and the Erotic Life of Property,* Lewis Hyde shows how gift exchange works to tie communities together. The natural world too is the locus of gift giving. While we may conceive nature as a brute and perhaps malevolent force to be conquered, as dead matter to be extracted, or even as virgin female stuff to be exploited, we also think of the natural world as inexhaustible abundance, the source of a bounty to be received and to be grateful for. The notion of nature's gifts bears deeper examination, for it illuminates not only our perceptual intertwinement with the natural world but also the way tradition and ritual can sponsor emerging newness.

In indigenous cultures, and in the remnants of Western religions, gratitude for nature's gifts may be expressed in first fruits ceremonies. A bit of the gift is returned to the land or the gods. The return gift to nature is thought to preserve nature's abundance. This is no token homage, no superstition. Increase belongs both to the essence of gift giving and to the essence of life: "The gift grows because living things grow."[70] The gift of life, reciprocated, means that we participate more fully in the cycle of life as such. The gift is multiplied because we in turn become more able gift givers. This is literally true in the first fruits ceremonies, where life's abundance and its renewal are the focus, and where our gifts indicate our awareness of the natural context that produces those fruits. "The forest's abundance is in fact a consequence of man's treating its wealth as a gift," Hyde notes.[71] If communities are maintained through the circulation of gifts, so too is the community of species. If, as Abram suggests, we and the beings of the natural world reciprocally create our perceptual capacities through our intertwinement with one another, then this cocreation is a reciprocal gift giving.

Gifts, by their very nature, must come from a source beyond our control. Otherwise they are not gifts but things taken under duress. The gift comes from the other, and it comes spontaneously and unexpectedly, taking us beyond ourselves into the new. In the very act of nourishing us, the gift received changes us. Our perceptual intertwinement with the natural world is a gift of this sort. Our perceptions come from beyond us, they are given to us unexpectedly, and they shape our perceptual capacities in turn. We are no longer what we were.

Just as for Ortega y Gasset the wild animal is the only true animal, so the wild is the truest gift. For as Jack Turner points out, the essence of the wild is that it cannot be grasped. It is that which is beyond us and beyond our control. Once we start managing wildness, the wildness is gone. The perceptions that come to us, gathered for us by the creatures we perceive, are wild in that sense. What we manage and control cannot nourish us with newness; it is like a "gift" we have arranged for someone to give to us.

How then can we understand agricultural communities, who certainly in some sense manage their land, to be receiving the gifts of nature? In what sense was Freeman House, a commercial fisherman at that time, receiving a gift from the wild when he took the heart of the dying salmon from its body and put it into his mouth? "The forest's abundance is . . . a consequence of man's treating its wealth as a gift," said Hyde. To some degree, certainly, it is *construing* our human activity as open-minded preparedness to receive—the hunter's attitude of alertness, in the form appropriate to the context—that makes the difference. The surprising spontaneity of his action is the sign of House's open alertness in that experience. That is the context of wildness that made his taking a receiving.

If the natural world gives to us through its ungraspable wildness, and if our perceptual capacities develop as a result of what we receive from it, then there will always be more wildness, more potentially to receive though we cannot know in advance what the gift might be. That is the sense in which nature's bounty is genuinely inexhaustible. In this sense, too, to receive the gifts of the natural world is to learn from experience and to be catastrophically changed in the process. Freeman House spontaneously received the gift of the salmon heart and experienced something beyond himself that changed him radically. Tangling at the roots of being is a potent process.

Nature's Bounty and the Labor of Gratitude

We stand before a bonfire or even a burning house and feel the odd release it brings, as if the trees could give the sun return for what enters them through the leaf.
—Lewis Hyde, *The Gift*

How are we to understand gratitude and reciprocity here? In what sense do we give something back to the natural world? Hyde introduces the notion of the labor of gratitude. The artist, the prototypically "gifted" person, serves as his example. Artistic talent is a gift; it has been given to the artist. But the artist must also earn this gift through the work of learning how to use it. This is the labor of gratitude.[72] Hyde retells the story of the shoemaker and the elves. The shoemaker and his wife are poor. The shoemaker has only enough leather for one more pair of shoes. He lays it out at night, and during the night naked elves come and make a beautiful pair of shoes out of it. This brings the shoemaker enough to buy leather for two pairs of shoes. Again he lays out the leather, and again the elves make him beautiful shoes from it. Eventually the shoemaker and his wife become curious enough to stay up at night and see what is going on. They see the elves at work, naked, and the next morning they decide to make them clothing—and shoes. Upon receiving this return gift, the elves depart, but the shoemaker retains the capacity he has gained from them and prospers. Making the return gift is his labor of gratitude, and when that labor is done he fully "owns" the gift he has received.

Hyde's tale is thought provoking. What is its analogue in the natural world? Remember the Papago hunter who pleases the supernaturals by his skill. In his dreams he hears singing, and in this way he acquires the songs of the ocean birds. What we might say here is that the work of hunting carefully is in fact the labor of gratitude for the abundance of the prey, and the song the hunter receives is at the same time the final outcome of that labor. Just as the shoemaker now can make beautiful shoes himself, so the hunter now "owns" the song.

The labor of gratitude links the work of tradition and ritual with the work of opening one's mind and heart. Through tradition and ritual we not only pass on skills and techniques, as for hunting or painting; we also pass on the means for opening one's mind and senses. To labor at acquiring those skills and that openness is the labor of gratitude. What one then receives through one's capacity for openness can be understood as the end of that process. When the hunter is given the gift of the prey's life, he has made the hunting process his own. That is the mark of the completion of the process. Giving the first fruits, or the one bit, back to the land or the gods then serves as an acknowledgment that the process has been completed.

To think of tradition and ritual as a labor of gratitude rather than preparation for receiving a gift is only a different way of marking the beginning and ending points in the ongoing process of giving and receiving that is our inter-twinement with the natural world. It emphasizes the reciprocal relationship of effort and openness. The hunter is grateful in advance, so to speak. Unless he has confidence that the gift of the prey will be received, he cannot sustain the broad sweep of openness that will allow the game to surge up unexpectedly into his field of vision. Sour, envious hunters are rare in flourishing indigenous cultures. They are people acknowledged to be in need of help.

The notion of the labor of gratitude also emphasizes the way in which recip-rocal gift giving with the natural world moves between the sensuous and the mental or psychic. For the "final" gift that the hunter receives is not the meat it-self but the dreams that he can then put to use in his life in the community. Such gifts can also be understood as the way that the natural world acquires consciousness. Cézanne works to acquire the skills of painting by absorbing the tradition of the Old Masters, and uses them to then open himself to the mountain and receive the internal experience of the mountain making itself visi-ble. What he receives is expressed in his paintings—that is, in his capacity to use his artistic talents and to give his paintings to the world where they can mingle with the minds of others. He now owns his talents and can give to the human community. But the landscape has made itself conscious in him, and in this sense Cézanne's painting is also his gift to the mountain that he has met and tangled with at the roots of being. Similarly, the hunter who becomes able to allow the game to surge into his field of awareness not only acquires dreams that mark his attainment and that serve his community; he has also been the instrument through which the game can become not only meat but also song. We use, and are used by, our gifts and our giving.

We find another version of the notion of the labor of gratitude in the words of the Sioux elder quoted earlier. Here the labor of gratitude takes the form of making oneself worthy: "If men would pay more attention to these preferences and seek what is best to do to make themselves worthy of that toward which they are attracted, they might have dreams which would purify their lives."[73] Here we might understand the initial gift to be the connection with the be-loved special animal, tree, plant, or spot of earth. When one makes oneself worthy, the gift changes form and becomes dreams that can then be further used to "purify lives"—that is, given outward again into the world.

Tangling at the roots of being is a meeting of essences. The prey gives the hunter the gift of its essence, expressed in the meat and received in the psyche as well as the stomach. The mountain gives Cézanne the gift of its essence, expressed in vision and received in the awareness of the very nature of vision. Receptivity is itself a gift, the gift of the open mind. It is developed through the effort contained in the instructions of tradition and ritual, but that effort only allows one to be open to the new. The gift back to nature, then, is the effort to make ourselves capable of perceptual intertwinement.

In a time when our presence has become so deadly, as Thomas Berry says, something more than a conventionalized ritual of first fruits is required. Making ourselves worthy may mean being willing to suffer the catastrophic change initiated by opening our perceptual capacities to the gifts of still deeper perception. The sacrifice called first fruits may mean the sacrifice of an old form of the self, the old self that is cut off from reciprocal perception and the cycles of life. Shifting our sense of self to include a broader identification with the whole of life in its cyclical form means not only an increase in love and caring, but also an increase in concern—and the loss, disequilibrium, and suffering entailed in such a deep psychic change. I think we must acknowledge this suffering and the anxiety it causes if we are to appreciate our profound resistance to the very possibility of a deeper intertwinement with the natural world.

Notes

1. See Donald Meltzer and Meg Harris Williams, *The Apprehension of Beauty*, esp. chap. 4, "On Aesthetic Reciprocity." For a fuller exposition of these ideas, see chapter 4 of the present volume.

2. Gary Snyder, "The Real Work," in *The Gary Snyder Reader*, 103.

3. Meltzer, *Studies in Extended Metapsychology*, 207–208, and see chapter 4 of the present volume.

4. See David Abram's discussion of "the enveloping earth" in *The Spell of the Sensuous*, for instance 203–204.

5. Snyder, "Kitkitdizze: A Note in the Net," in *The Gary Snyder Reader*, 283.

6. See Paul Shepard, *The Tender Carnivore and the Sacred Game*, 160–169.

7. See Meltzer et al., *Explorations in Autism*, esp. chap. 2.

8. Maurice Merleau-Ponty's discussions of Cézanne are found in "Cézanne's Doubt," in *Sense and Non-Sense*, and "Eye and Mind," in *The Primacy of Perception*.

9. Merleau-Ponty, "Cézanne's Doubt," in *Sense and Non-Sense*, 17.

10. Merleau-Ponty, "Cézanne's Doubt," in *Sense and Non-Sense*, 12.

11. For an illuminating comparison of Cézanne's paintings with photographs of the places he painted, see Erle Loran, *Cézanne's Composition: Analysis of His Form, with Diagrams and Photographs of His Motifs* (Berkeley: University of California Press, [1943] 1985).

12. Hemingway studied the Cézanne paintings he saw in Paris in the 1920s. Note his later statement: "I can make a landscape like Mr. Paul Cézanne," cited in Lillian Ross, *Portrait of Hemingway,* 60.

13. Merleau-Ponty, "Eye and Mind," in *The Primacy of Perception,* 166.

14. "Eye and Mind," 163. For a full discussion of vision as relational, drawing on Merleau-Ponty, see Laura Sewall, *Sight and Sensibility: The Ecopsychology of Perception.*

15. Merleau-Ponty, "Eye and Mind," in *The Primacy of Perception,* 163.

16. Merleau-Ponty, "Eye and Mind," in *The Primacy of Perception,* 164.

17. Merleau-Ponty, "Eye and Mind," in *The Primacy of Perception,* 165.

18. Merleau-Ponty, "Eye and Mind," in *The Primacy of Perception,* 167.

19. Merleau-Ponty, "Eye and Mind," in *The Primacy of Perception,* 166.

20. José Ortega y Gasset, *Meditations on Hunting,* 150–151.

21. Ortega y Gasset, *Meditations on Hunting,* 150.

22. Ortega y Gasset, *Meditations on Hunting,* 139–140.

23. Ortega y Gasset, *Meditations on Hunting,* 141.

24. Ortega y Gasset, *Meditations on Hunting,* 140.

25. Ortega y Gasset, *Meditations on Hunting,* 142.

26. Ortega y Gasset, *Meditations on Hunting,* 142.

27. Ortega y Gasset, *Meditations on Hunting,* 150–151.

28. Snyder, "Kitkitdizze," in *The Gary Snyder Reader,* 283.

29. Shepard, *The Tender Carnivore and the Sacred Game,* 153.

30. Shepard, *The Tender Carnivore and the Sacred Game,* 154.

31. Ruth Underhill, *Singing for Power* (Berkeley: University of California Press, 1968), quoted in Snyder, "The Politics of Ethnopoetics," in *A Place in Space,* 144.

32. Ortega y Gasset, *Meditations on Hunting,* 152.

33. See Wilfred Bion, "Notes on Memory and Desire," in *Cogitations,* 380–385.

34. Ortega y Gasset, *Meditations on Hunting,* 150.

35. "The torch of chaos and doubt—this is what the sage steers by" (*The Basic Writings of*

Chuang Tzu, trans. Burton Watson, 37). For an extrapolation in the field of contemporary culture and education, see Guy Burneko, *By the Torch of Chaos and Doubt.*

36. Quoted in James Hillman, *The Thought of the Heart and the Soul of the World,* 87–88.

37. Hugh Brody, *Maps and Dreams,* 43, quoted in Morris Berman, *Wandering God,* 8. Although not the central subject of his book, Berman's discussion of hunting-gathering consciousness as a horizontal, paradox-oriented culture resonates with my discussion of a "field" consciousness. "In this world," writes Berman, "the secular is the sacred, which is all around us. This is why I call it a horizontal perception" (p. 11).

38. Abram, *The Spell of the Sensuous,* ix.

39. Abram, *The Spell of the Sensuous,* 62.

40. See Bion, "A Theory of Thinking," in *Second Thoughts,* 119.

41. See Abram, *The Spell of the Sensuous,* 61–62.

42. Abram, *The Spell of the Sensuous,* 61–62.

43. See Bion, "Making the Best of a Bad Job," in *Clinical Seminars and Other Works,* 321.

44. Abram, *The Spell of the Sensuous,* 63.

45. Abram, *The Spell of the Sensuous,* 66–64.

46. Abram, *The Spell of the Sensuous,* 64.

47. James Hillman, *The Thought of the Heart and the Soul of the World,* 111–112.

48. Gary Paul Nabhan, "The Far Outside," in David Clarke Burks, ed., *Place of the Wild,* 23.

49. Freeman House, *Totem Salmon,* 112.

50. House, *Totem Salmon,* 112–113.

51. House, *Totem Salmon,* 113.

52. House, *Totem Salmon,* 113.

53. Paul Shepard, "Nature and Madness," in Roszak, Gomes, and Kanner, eds., *Ecopsychology,* 37.

54. Shepard, "Nature and Madness," in Roszak, Gomes, and Kanner, eds., *Ecopsychology,* 27.

55. Shepard, "Nature and Madness," in Roszak, Gomes, and Kanner, eds., *Ecopsychology,* 30.

56. Shepard, "Nature and Madness," in Roszak, Gomes, and Kanner, eds., *Ecopsychology,* 37.

57. Shepard, "Nature and Madness," in Roszak, Gomes, and Kanner, eds., *Ecopsychology,* 37.

58. See Merleau-Ponty, "Cézanne's Doubt," in *Sense and Non-Sense,* 19.

59. Merleau-Ponty, "Cézanne's Doubt," in *Sense and Non-Sense,* 21.

60. Merleau-Ponty, "Cézanne's Doubt," in *Sense and Non-Sense,* 19.

61. Merleau-Ponty, "Cézanne's Doubt," in *Sense and Non-Sense,* 19.

62. Merleau-Ponty, "Cézanne's Doubt," in *Sense and Non-Sense,* 20.

63. Merleau-Ponty, "Cézanne's Doubt," in *Sense and Non-Sense,* 17.

64. Merleau-Ponty, "Cézanne's Doubt," in *Sense and Non-Sense,* 17.

65. See Bion, "On a Quotation from Freud," in *Clinical Seminars and Other Works,* 307.

66. See Adrian Stokes, *Painting and the Inner World,* 7.

67. Merleau-Ponty, "Cézanne's Doubt," in *Sense and Non-Sense,* 17.

68. See Shepard, "On the Significance of Being Shaped by the Past," in *The Only World We've Got,* esp. 136. In this context, see also Thomas Berry's discussion of the Native Americans' capacity to survive and endure over the past five centuries of European contact, and their interior sources of renewal, in "The Historical Role of the American Indian," in *Dream of the Earth,* 180–193. Surely these capacities are related to earlier experience within a stable— that is, only slowly changing—natural and cultural framework.

69. Shepard, "Nature and Madness," in Roszak, Gomes, and Kanner, eds., *Ecopsychology,* 39.

70. Lewis Hyde, *The Gift,* 27.

71. Hyde, *The Gift,* 27.

72. See Hyde, *The Gift,* chap. 3, "The Labor of Gratitude." Hyde's retelling of the story of the shoemaker and the elves is on pp. 48–50.

73. House, *Totem Salmon,* 54.

4   What Beauty Can Tell Us: The Face of Nature

What we naively call beauty, and what we experience as artistic force . . .
—Christopher Alexander, *A Foreshadowing of 21st Century Art*

The experience of beauty epitomizes the intimacy of reciprocal perception. The beautiful is seen within a reverberating perceptual field that it shares with the Other. In other words, what we perceive as beautiful is experienced as having a face that returns our gaze. Aesthetic experience—the experience of beauty—gives rise to this field that holds those who gaze at one another. Beauty is common to both the human and the natural worlds. What we naively call beauty is our experience of the felt presence of the "ten thousand things" ablaze in the field of the universe. The flame of imagination is the hunter's alertness under another name.

Aesthetic Reciprocity

The baby must be held as an aesthetic object by the mother for the experience of their love-making to reverberate and escalate in intensity.
—Donald Meltzer

In the mutual gaze of mother and baby we see the prototypical instance of aesthetic reciprocity. Aesthetic experience begins at birth, if not before, and is the experience that links the baby with the mother. That experience becomes the basis for curiosity, interest, and the exploration that later extends out to the world. At first, however, the mother *is* the world: "In the beginning was the aesthetic object, and the aesthetic object was the breast, and the breast was the world,"[1] writes Donald Meltzer. The baby, Meltzer says, perceives his "ordinary beautiful devoted mother" as an object of beauty, both physical and mental. The overwhelmingly interesting sight of her breast and her face, her nipples and her eyes, "bombards him with an emotional experience of a passionate quality, the result of his being able to see these objects as 'beautiful.'"

Mystery is an important part of the baby's initial experience of his beautiful mother. The mother's psyche, as expressed in her emotions and her actions, inspires as passionate a response in the baby as her physical beauty, and is even more mysterious: "The meaning of his mother's behavior, of the appearance and disappearance of the breast and of the light in her eyes, of a face over which emotions pass like the shadows of clouds over the landscape, are unknown to him."[2]

The enigmatic quality of the mother's beauty gives rise to what Meltzer calls the "aesthetic conflict"—the conflict between passionate attraction to, and anxious suspicion of, the powerful unknown that the mother's beauty represents.[3] In the grips of the aesthetic conflict, the baby cannot tell whether the mother is a Beatrice or a Belle Dame sans Merci. Remember Meltzer's story in chapter 1 about the young boy who was amazed by an experience that did not immediately yield its meaning. The newborn responds to the mother's overwhelming beauty with similar amazement. It is as though the infant has suffered what Joseph Campbell calls "aesthetic arrest"[4] and is transfixed by everything in sight. This experience of beauty is not yet understandable. Hence it is overwhelming and, to some degree at least, unbearable. But at the same time it inspires development and growth. The child wants to be able to grasp and tolerate this experience in order to fully know the mother and to receive the gift her beauty promises. As Meltzer puts it, "What the fragile ego of the child cannot sustain and is riven by, a lifetime of development strives to restore so that the beauty of the object may be looked upon directly without 'damage to the soul,' as Socrates feared."[5]

Meltzer's notion of the aesthetic conflict as a spur to growth and development is a clue to understanding the tremendous maturity the natural world requires of us. If the experience of aesthetic reciprocity is the foundation for our capacity to enter into an experience of perceptual reciprocity with the natural world as well,[6] then what works against our capacity to tolerate aesthetic experience will work against our capacity to relate to the natural world as kin.

The mutuality of the aesthetic experience is crucial to the process of learning to endure the aesthetic conflict so that we can look upon beauty. The beautiful mother is impacted in turn by the beauty of her baby. As we have seen, the mother's reverie—the mental state in which she "holds" or "contains" the baby's as yet inchoate psychic experience and ponders its meaning—makes that experience less overwhelming to the baby. Like a kind of psychic nourishment, the baby begins to take in the mother's capacity to ponder the meaning of experience. But it is not simply that the mother "holds" the baby in her reverie. When mother and baby gaze at one another with mutual passion and experience the aesthetic impact of one another, they are in effect in a "containing" space together. The mutuality creates a shared field. Within that field, Meltzer tells us, the "experience of their love-making" can "reverberate and escalate in intensity."[7] Here is that abyss of infinite escalating impact that results in amazement, the aesthetic arrest. But here too is the safety, the shared field within which the boundlessness of aesthetic experience can become tolerable.

When the mother does not see the baby as beautiful—that is, does not respond to the baby as an aesthetic object—the consequences can be devastating. Meltzer tells about the case of little Claudia, born deformed to an elegant, beautiful mother.[8] Claudia experienced herself as an "ugly little clown." She could not conceive how babies were created (though she knew the facts of the matter), for she could not conceive an experience of love making that would give rise to an ugly little clown like herself. Nor, without being an aesthetic object for her mother, could she experience the mother-baby relationship as an experience of love making. A field of perceptual reciprocity, a creative coupling of two beings, seemed an impossibility to her.

The mother's face is the original locus of aesthetic experience, and the prototypical form of aesthetic reciprocity is a mutual gazing into one another's eyes and at one another's faces. The face is an object utterly pregnant with significance. It "flashes the sign of presence," writes Michael Eigen.[9] To see the face is to be in the other's presence. It is through the face that we enter into reciprocity with another being, because through the Other's face a separate being turns toward us and returns our gaze. The face connotes relatedness because in the mutuality of gazing we are together in the aesthetic field. At the same time, it connotes separateness, not only because it faces—that is, confronts—us, but also because it is unique. No two faces are the same. Thus, the experience of facing one another establishes kinship without identity. In its turning and gazing, the face expresses intentionality—an interior subjectivity that shines out in the exterior features, making the beautiful face itself, as Meltzer tells us, a confluence of matter and spirit, interior and exterior.

Donald Winnicott provides a further perspective on the mutual gazing of mother and baby. Winnicott poses the question, What does the baby see when the baby looks at the mother's face? The mother's face is a mirror, is his answer, and the baby sees himself or herself: "Ordinarily, what the baby sees is himself or herself. In other words, the mother is looking at the baby and what she looks like is related to what she sees there."[10] It is this experience of being "seen" by the mother, reflected in her responses, as it were, that first gives the baby a sense of self. Perhaps the mother's face reflects the baby's emotion; perhaps her delight and surprise mirror the baby's sense of discovery. If, conversely, the mother's face is a fixed or a self-absorbed mirror, the baby will see the mother, but will not have a sense of his or her own self in rapport with the mother. The baby will experience the nonmirroring mother as something external in a way that deprives the baby of connection with her.

In the experience of being seen by the "mirror-mother," Winnicott says, the baby establishes a sense of self-in-relationship, a sense of difference within union. Being seen in this way provides the baby with the sense of being in the reverberating field of aesthetic reciprocity. This sense of kinship, of difference-in-unity, is the basis of the baby's capacity for what Winnicott calls *creative looking*.[11] It is the basis for the baby's confidence in his or her ability to find meaning in the world of seen things—to perceive aesthetically. At first the mother's face is the baby's world. Then the world is extended, and it too is a face. Without this initial sense of self-in-relationship, the other is not experienced as a full person but as something more like an external object, an "it," as in Martin Buber's distinction between an I-Thou relationship and an I-it relationship. Without this sense of kinship, the world too is perceived as not-kin and without the infinite meaningfulness of an aesthetic object. Meaning comes through emotional connection, not through the detachment experienced toward something that is merely an "it."

Winnicott tells about a patient who was a sensitive judge of the visual arts. Her early life had been very disturbed in that she was not seen in the way that gives one a sense of self-in-relationship. As an adult, a lack of beauty in the external environment had the effect of disintegrating her sense of self. She could distinguish lack of beauty by the fact that she felt awful.[12] Winnicott implies that his patient might not have been so profoundly affected by the aesthetic state of the external world if her early life had not been so disturbed. Perhaps so; but I think there is another equally important lesson here. The capacity for aesthetic experience is necessarily hard won and difficult to maintain, both because of the mother's inevitable failures, however minor, to participate in aesthetic reciprocity with the baby, and because of the terrors and suspicions of the unknown that engender what Meltzer calls the aesthetic conflict. The human-made environment to which Winnicott's patient was so sensitive reflects our own aesthetic incapacities, which in turn affect our sense of self. There is a reciprocity of ugliness and failure of relationship as well as an aesthetic reciprocity of reverberating love making. As Paul Shepard has remarked, "Life in a degraded environment . . . doesn't make you feel at home."[13]

The Work of Art as Mirroring Face

It is as if the thing, the bronze, or the carpet, establishes itself in my own belly, as a voice, speaks with my own voice, exists with my own force, and forces my awareness of an ultimate mother, or an ultimate creature of which I am a part—and which exists in me.
—Christopher Alexander, *A Foreshadowing of 21st Century Art*

How amazing to think, Christopher Alexander once remarked, that a made object might be of such a nature that it could reveal the core of the universe itself.[14] To ask how the made object can reveal the core of the universe is to ask how the nonhuman can have a face. It is also to ask how we can be kin to the nonhuman, thus to the natural world. Does Nature have a face? The work of art, the epitome of the beautiful made object, reminds us that we can experience aesthetic reciprocity with the nonhuman.

One way the made object can reveal the nature of the universe is that it mirrors the human self. That is how we know that it is a work of beauty—we feel a commonality of self with it. Winnicott's patient found a mirror of nonself, a disintegrated self that was not in relationship, in an ugly environment, just as mass-produced objects without variation reflect our lack of capacity for creative interchange with the nonhuman. When the made object is beautiful, however, it reflects back to us our own capacities for creative looking.

The beautiful work makes him aware of an "ultimate mother," Alexander says.[15] As babies see their own aesthetic capacities and their own beauty reflected in their first aesthetic object, the mother's face, so we experience the beautiful work of art too as a mirroring face into which we look with fascination. If aesthetic reciprocity takes the form of the field of love between mother and baby, the work of art, as Adrian Stokes says, is "an emblem of the state of being in love."[16]

Artists know that nonhuman things can have faces and enter into reciprocal interaction with us. Paul Klee claimed that at a certain point the painting acquired a face. "Now it looks at me," he said. The looking that we speak of in aesthetic reciprocity, as we know, is really a multisensory, indeed synesthetic activity. I don't play music, Bach is reported to have said, it plays me. Theodor Adorno extended this idea to say that we don't understand music, music understands us. Music is not visible, but it can recognize us and express itself to us, as a face would.

It is not the work's subject matter that makes it a mirroring face. The work—say a painting—might have any subject matter or be completely abstract. Adrian Stokes, painter and writer on art and aesthetics, compares the work of art to a face in that like a face it condenses emotion into a sensory form. The contained physical object manifests something that goes beyond itself. He speaks of the work's "flowering radiant compactness." Like the mother's face, the work smiles at us, showing us its "beneficence in space" and inviting us to an exchange by way of what Stokes calls its "incantatory quality."[17]

"The Invitation in Art," Stokes called one of his essays. The work reaches out to us and draws us in. But what does the work invite us into? Like a face, the work seems to invite us to make contact with the subjectivity it expresses. The work invites us to mingle our own subjectivity with the psychic contents the artist has embedded in its sensory forms. Though made of sensory stuff, it has a psychic reality of its own that can meet and be met by the viewer's own psyche in what Stokes calls "a passionate confluence between the forms of the inner self and those of the aesthetic object."[18]

How does a sensory object accomplish this confluence of inner, psychic realities? It is the processes laid down in the work's form, Stokes says, that link our subjectivity with the work's. The work's physical features—the movement of the brush strokes or the complexity of color patterning in a painting—are experienced by the viewer as analogues of corporeal processes with their attendant imagined responses: activities, say, of exploring, attacking, compressing, merging.[19] The confluence between the sensory experience and the imagined bodily interaction gives the sense of merging. At the same time, the work maintains its autonomy, its separate wholeness, which is achieved through the closure of artistic form. In this way, as Stokes says, the work of art is an emblem of the state of being in love: we experience merger with the beloved while remaining aware that we are separate individual beings.

Just as the baby is not sure whether the beautiful mother is Beatrice or the Belle Dame sans Merci, art's invitation entails risk. We do not know beforehand all that we will meet and mingle with in this merger. The work, like the face, promises infinite and unknown subjective contents, which we can only know within the experience of aesthetic reciprocity. "Mystery speaks through [art's] objective form," say Meltzer and his coauthor Meg Harris Williams, "but it can only be observed by the viewer who has surrendered his inner self for observation."[20] Aesthetic experience is not a matter of looking at an object; the look is reciprocal, and it is intimate. Faces, including those of works of art, look back at us, and like the mother they are capable of more than one expression. The work may extend its invitation to intimacy with a benevolent smile, but it contains negative or destructive psychic forces as well. A work that does not, has split itself off from the whole of the universe and truncated itself. Thus intimacy with the work of art, like intimacy with a beloved person, means opening oneself to engagement with destructive forces as well as constructive ones, with negativity as well as beneficence.

In all of this, however, Stokes insists, a sense of beauty, bountifulness, and integration prevails. The integration accomplished by aesthetic form—the integration through which the work becomes autonomous, whole, and separate—takes in and binds the negative or destructive forces. "The successful work has the strange power to teach its own lesson," said Merleau-Ponty. Viewers who surrender themselves to observation by the work, who resonate with the positive and the destructive forces in it and allow their psychic structures to be reshaped by the integrative powers of form, will have learned the work's lesson and received the work's bounty.

This is the sense in which the work of art, a made object, has a face. For the natural world to have a face in this sense would mean opening ourselves up to exploration and reshaping by it through the very act of responding to its invitation.

The Inviting Face of the World

> *The aesthetic sense perceives the form of things, apprehending the particular*
> *shape of each event, its nature disclosed by its face.*
> —James Hillman, *The Thought of the Heart and the Soul of the World*

James Hillman is the spokesperson for the soul of the world, the *anima mundi*,
and the invitation it extends to us through its sensuous beauty. "Beauty is an
epistemological necessity," says Hillman. "It is the way in which the Gods
touch our senses, reach the heart, and attract us into life."[21] Which god?
Aphrodite, of course, goddess of erotic love, the alluring goddess who draws us
to her through the beauty of form, the magic of appearance. Beauty is the quality of the sensory world—the "sensible" world, in David Abram's term—which
allows us to respond to it with what Hillman calls "the thought of the heart."
We respond to beauty with the ardent response of a lover.

For Hillman, this aesthetic response is a gasp of amazement at the beauty of
the object. In this gasp—this in-breath or inspiration—we perceive the particular individuality, the "eachness," in William James's term, of the thing we see,
and we enter into an interior relationship with it. The heart, our interior essence, touches the interior essence of the thing perceived. "The transfiguration
of matter occurs through wonder," Hillman says; "the aesthetic reaction which
precedes intellectual wonder inspires the given beyond itself, letting each thing
reveal its particular aspiration within a cosmic arrangement."[22]

The aesthetic response is thus a revelation of individuality, an individuality
inseparable from, in fact only perceived within the context of intimacy, a heart-
to-heart touching. The aesthetic response establishes an interior or imaginative
reciprocity in which individuality unfolds as it is perceived. As Hillman puts it,
when we take the aesthetic object in through the "thought of the heart," the
imaginative existence of the object is activated, and the object unfolds its essence—becoming more beautiful in the process. Deepening intimacy is deepening unfolding within the sphere of imagination, and intensifying beauty.[23]

"Perceptibility," Novalis said, "is a form of attention."[24] It is our attentive perceiving—what the ancients called "*notitia*, attention to the qualities of things,
the capacity to form true notions of things from attentive noticing"[25]—that sets
this process of unfolding in motion. The aphroditic allure of the beautiful object arouses our attention. We open our perceiving hearts to take in the loveli-

ness of the thing perceived, and it unfolds its loveliness, inspired in turn by our aesthetic response. The love making reverberates and escalates in intensity.

The Aesthetic Field: Aura and the Imaginal Dimension

> *If I have an interest in preservation, it is preserving the power of presence—*
> *of landscape, art, flora, and fauna. . . . One might suppose it is something that*
> *could be added on later. But no, . . . the loss of aura and presence is the main*
> *reason we are losing so much of the natural world.*
> —Jack Turner, *The Abstract Wild*

The reverberating intensity of the aesthetic response transpires through the activity of the imagination—what the English Romantic poets would call imagination, a creative capacity, as opposed to fantasy—and in the realm of what Hillman, following Henri Corbin, calls the *imaginal*. For the Persian mystics Corbin studied, the imaginal is intermediate between the material and the spiritual worlds and links those world with one another.[26] It is a quasi-sensuous but nonmaterial dimension, perceivable by the imagining heart, as Hillman might say. It is not a realm of daydreams or hallucinations but simply the real world and the real psyche contacting one another via the imaginative capacity, joined in what that contact creates and perceives.

In this imaginal dimension things appear in their essence, and things are linked through their resemblance to one another. The imaginal dimension is in fact the field of aesthetic reciprocity, the area in which the "confluence" of self and other, as Adrian Stokes would say, takes place. As Corbin explains it, the imagination of the seer changes the quality of the sensory data to allow their interior essence to be seen, just as Hillman says. The sensory data become "a pure mirror, a spiritual transparency," and the earth and its creatures are "raised to incandescence" by the imagination of the perceiver.[27] They are seen by the imagination in their "light of glory," or Xvarnah. In this light of glory, which is both the activity of the imagination and its medium, the soul's own light, the light of the things of the earth, and the imagination itself are reflected in and joined with one another. It is the medium of kinship. What appears in the light of glory remains flesh—spiritual flesh. What Corbin calls "the earth of visions" is, he explains, "the place where spirit and body are one, the space where spirit, taking on a body, becomes the *caro spiritualis,* 'spiritual corporeity.'"[28] Corbin's "spiritual flesh" is what Merleau-Ponty calls "the flesh of the world." As in Merleau-Ponty's conception, imaginal vision happens within and between beings by means of the light that is their living presence.

The "light of glory" is another name for the aura or halo, the diaphanous light that surrounds a living being and is the token of its imaginal existence. The aura has a central place in beauty. It is the quality of beauty's presence and it arises in the interchange between perceiver and perceived. If, as Novalis said, perceptibility is a form of attention, "the perceptibility [Novalis] has in mind," according to Walter Benjamin, "is none other than that of the aura. . . . To perceive the aura of an object we look at means to invest it with the ability to look at us in return."[29] Beauty's aura, in other words, is both the object and the product of a reciprocating attention.

"Beauty is the object of experience in a state of resemblance," wrote Benjamin, in a formulation that defines the aesthetic field. In the aesthetic field the participants are kin, joined by resemblance, and it is as kin that we perceive the aura. Benjamin's term for this aesthetic field, which we have called the imaginal dimension, is the *dreamworld*. Thus objects in dreams, invested with aura, return our gaze: "To say 'Here I see such and such an object' does not establish an equation between me and the object," Paul Valéry pointed out. But "in dreams . . . there is an equation. The things I see, see me just as much as I see them."[30] Benjamin calls this dreamworld a "universe of intertwinement," noting that reciprocity means resemblance as well: "The similarity of one thing to another which we are used to, which occupies us in a wakeful state, reflects only vaguely the deeper resemblance of the dreamworld in which everything that happens appears not in identical but in similar guise, opaquely similar one to another."[31]

Dreams, of course, are only one instance of the imaginal. It is the imaginal dimension as such, the dimension of aura, that is the locus of beauty and thus of kinship. Resemblance, we might say, is the face of kinship. The aura is its emanation.

In the aesthetic field, we are kin to, and can experience our kinship with the natural world as well. Beauty is not an accidental quality of particular places or plants or animals. It is the mutual experience of essence and individuality in the imaginal dimension, which is the medium of our aesthetic response. Our kinship with the natural world is with nature as the wild, as that which we meet in its interior of its essence and which cannot be grasped or controlled. Hence the wild too is invested with aura, the sign of its presence. As Jack

Turner says, the wild is destroyed when its aura is destroyed, for both depend on presence. "If I have an interest in preservation," Turner says, "it is preserving the power of presence. . . . The loss of aura and presence is the main reason we are losing so much of the natural world."[32]

The Flame of Life Emerging from the Void

This quivering and hurrying lambency of beings . . .
—J. B. Priestley

The earth is not only other creatures; it is also landforms, rocks, water, air. Once again, we are prompted to ask, Does Nature have a face? How are we to conceive the personhood of the nonsentient?

The imaginal dimension points us to the ultimate continuities and kinships between the human and the nonhuman worlds. As Corbin says, in the imaginal dimension the image of the earth and the image of the soul arise in conjunction with one another. The work of art too points the way to understanding the personhood of the nonsentient. The work of art is itself nonsentient, made of physical materials. Its human-infused meaning gives it its face and its continuity with our own psyches.

If we follow beauty back to its essential nature, then, it links us to life and selfhood and to all beings in a way that transforms our notion of space and matter—the nonsentient. Life and personhood become qualities of space and matter in a way that connects them to us. Inner and outer are closer than we might have thought. Reflection on creativity—the process whereby things come into being—illuminates these connections. For the Buddhists and for many mystical traditions, the void is seen as the ground from which beings arise. This notion of emergence from the void provides a way to look at the process of unfolding that is creativity. The void is the source of the "ten thousand things"—that is, all sentient and nonsentient beings. All emerge from the void as the basic ground of things, and all things lead us back to the void in turn. Form is emptiness; emptiness, form, say the Buddhists. Words are silence; silence, words, Gary Snyder might say. He writes of the "silent" Chinese nature poems: "The reverberation of mirroring—a flight of birds across the mind of the sky—leaves an afterimage to be savored, and finally leaves no trace."[33]

The light of glory called the aura is generated in this process. In a dream reported by J. B. Priestley this light becomes the flame of life emerging from and passing back into the void:

> I was standing at the top of a very high tower, alone, looking down upon myriads of birds flying in one direction, every kind of bird was there, all the birds in the world. It was a noble sight, this vast aerial river of birds. But now in some mysterious fashion the gear was changed, and time speeded up, so that I saw generations of birds, watched them break their shells, flutter into life, mate, weaken, falter, and die. . . . As I stared down, seeming to see every creature's ignoble little history almost at a glance, I felt sick at heart. It would be better if not one of them, not one of us, had been born, if the struggle ceased forever. . . . But now the gear was changed again, and time went faster still, and it was rushing by at such a rate that the birds could not show any movement, but were like an enormous plain sown with feathers. But along this plain, flickering through the bodies themselves, there now passed a sort of white flame, trembling, dancing, then hurrying on; and as soon as I saw it I knew that this white flame was life itself, the very quintessence of being. And then it came to me, in a rocket-burst of ecstasy, that nothing mattered, nothing could ever matter, because nothing else was real but this quivering and hurrying lambency of beings.[34]

This dream portrays the development of perception to the point where the white flame of life, "this quivering and hurrying lambency of beings," can be seen to be the source and nonmaterial essence of the ten thousand things, the beings that live and die as individuals. I will call this the process of *perceptual maturation*. One must go through a process of encompassing life and death in one's perception, and experiencing the suffering entailed, to perceive the life flame as the trumpeting of the void. As Dogen said: "To forget yourself is to be authenticated by all things. To be authenticated by all beings is to effect the molting of the body-mind, both yours and others."[35]

Beethoven's late work, which, as Charles Rosen has remarked, is "essentially meditative,"[36] offers a musical experience of such perceptual maturation. The void makes its presence deeply felt in the fully individualized variation sets in Beethoven's final piano sonatas, where there is a simplicity of underlying melody and tonality that unfolds rhythmically and harmonically in coexisting voices moving at different speeds. The Arietta, the slow movement of Beethoven's last piano sonata, op. 111, for instance, is a set of graded variations in C major. In the final variation a trill in the fastest moving voice "succeeds as almost no other work," Rosen says, "in suspending the passage of time."[37] The trill takes on the quality of Priestley's lambent quivering life flame, while at the same time one hears the more substantive "layers" of movement reminiscent of the flights of individual birds or the plain of feathers in Priestley's dream. Life as the life flame emerges from the silence and timelessness of the void, and gives rise to all the forms in between. Beethoven too is intent on teaching us that lesson.

If the void gives rise to all things, the distinction between sentient and nonsentient beings ceases to be relevant. "Blue mountains are neither sentient nor insentient," says Dogen in the *Rivers and Mountains Sutra.* "You," he continues, "are neither sentient nor insentient. At this moment, you cannot doubt the blue mountains walking."[38] A mountain is a life form in itself, the rocks that form it arising and decaying at their own speed. At the same time it is composed of the myriad beings, sentient and nonsentient, for which it is the dwelling place. What dwells within the mountain is part of the mountain. The notion of "thinking like a mountain," taken up by the Deep Ecologists,[39] reminds us that the usual sharp divide between sentience and nonsentience is superseded by a broader conception of life forms.

Cézanne, as we have seen, thought like a mountain, and the mountain thought itself in him. There is a strange phenomenon in his paintings that I call "the navel." It is a small area of darkness, an enclosed concentrated point somewhere in the painting, quite inconspicuous, which seems to link the painting back into a dimension of dark emptiness.[40] The objects in the painting seem, as I said earlier, to form a quasi-solid substance, so cohesive are their interrelationships, and so fully do they enclose the spectator in their containing space—but those very objects in their solidity seem to have emerged from the darkness "behind" them. The navel is their point of emergence. Not only does this linkage speak to the meeting of nature and mind in the interior of Cézanne's seeing consciousness, it also suggests an intertwining of interior and exterior, matter and spirit, in the objects themselves. It is as though the paintings are showing us the emergence of the ten thousand things from the void.

To think of all things emerging from the void requires a shift in our notion of what it means to dwell somewhere. From the perspective of the ten thousand things emerging from the void, home and homelessness become identical in the sense that every place has the same origin and every place is a possible place to recognize the common ground in the void. Hence "a mountain always practices in every place."[41] Put in Corbin's terms, if we are grounded in earth, we are homeless in the sense that we have no one limited home but rather meet the earth in the imaginal dimension. In the same vein, Snyder reminds us that the Buddhist word for priest or monk also means "homeless": "Homelessness," he comments, "is here coming to mean 'being at home in the universe.' . . . The condition of true 'homelessness' is the maturity of relying on nothing and responding to whatever turns up on the doorstep."[42] It is as though one needs to be at home in the field of the whole in order to be open to what emerges. From this point of view, the wanderings of hunting-gathering societies mark the hunter's willingness to be in the field wherever it is.

At the same time, the mind willing to receive what emerges receives it in its utter specificity. Each of the ten thousand things is individual. Each reception is a locus of home. This notion of homelessness, then, extends the notion of being at home to the larger cosmic connections between earth and soul that arise through what we might call not only perceptual but also imaginal maturation.

Beauty and the Darkness of Death

Art has two constant and unending preoccupations; it is always meditating upon death, and it is thereby creating life.
—Giuseppe Tomasi di Lampedusa

We think of beauty as radiant and luminous—self-luminous, perhaps—and the essence of life or spirit as blazing light, as in Priestley's dream-vision of the flame of life hurrying along the plain of feathers. But the ultimate source of life, the fertile void, also holds darkness, in fact blackness, as in the Persian mystics' "black sun."[43] For, as the Jungians would say, the ultimate source encompasses opposites. In Christopher Alexander's words, "There is at the heart of all things, a single voice of universal blackness and thickness and light, that speaks in all tongues and holds all force in itself."[44]

Certainly the natural world, the epitome of life, confronts us with death and decay, which are inevitably conceived and experienced in terms of darkness, however fertile the darkness may ultimately be understood to be. Though we are reluctant to confront them, death and decay are part of nature's principle of operation. "Life in the wild is not just eating berries in the sunlight," Gary Snyder admonishes us. Reminding us that wild systems can be seen as "irrational, moldy, cruel, parasitic," he evokes a "'depth ecology' that would go to the dark side of nature—the ball of crunched bones in a scat, the feathers in the snow, the tales of insatiable appetite."[45]

The dark side of nature is an interior experience as well, a "darkening of the light" in which death and deathfulness are experienced not only as the death of another creature but also as a darkening within the psyche that somehow sucks the light and life from one's experience of the world. The hunter's alert mind, open to the emergence of the prey in the field of vision, requires him to be fully open to the experience of death as well. In his memoir *Yet Being Someone Other,* Laurens van der Post records a conversation with a man named Kaspersen, captain of a whaling boat on which van der Post was to ship. Kaspersen asked him whether he had ever shot an elephant, and van der Post replied that he had. How did he feel afterward, Kaspersen asked. "I hate myself for it," van der Post replied, "not only at the time but for days afterwards." "Why?" asked Kaspersen. "Because," replied van der Post, "at one moment there is so much life through the sights of my gun; the next I pull the trigger and there's nothing at all. It's as if a great black hole had been made in the day—and everything around suddenly goes dark and empty."[46]

Kathleen Raine describes a still more interior experience of deathfulness as darkness. A young cousin of hers died. The boy, to whom she had not been particularly close, had suddenly said to the friend he was walking with, "It is so dark and cold on this side of the road, let us cross into the sun." He crossed the road, collapsed, and died two days later of meningitis. Raine then experienced an invasion of her own consciousness by death: "The horror that took possession of me was, literally, a darkening of the light. The light was filled with blackness, the light itself was emptied of light; not figuratively, but in a way those only who have known the like will understand, the phenomenal world ceased to be accessible to me. For me the bright Easter sun was dark, shedding blackness."[47]

The sun shed blackness—an image of the ultimate darkening of the light, experienced as an alienation from all that is real. Raine describes seeing golden dandelions in the sun, filled with blackness, "emptied of all substantial reality." This sense of darkness at the heart of the phenomenal world is an experience of terrible aloneness. There is no feeling of psychic or emotional connection with other beings. "I could see objects around me, and people, but I could not reach them," writes Raine. "Between them and me there was a great gulf fixed."[48] If life is the experience of being separate, differentiated, within connection, then this experience of deathfulness is the experience of being absolutely separated, severed from all connection.

If darkness and light are both encompassed in the ultimate source, the void, how are we to understand death and destruction in their relationship to life and wholeness? And how are we to understand the fact that the natural world, the experience of beauty, and the emotional connection we call love, all fill us with the conviction that life is greater than death? Perhaps it is the effort to penetrate the dimension of darkness and death with some kind of consciousness that helps us to incorporate that dimension within the larger whole that is both a fertile void and a luminous ground.

By the effort to penetrate the darkness with consciousness, however, I do not mean an attempt to obliterate the darkness with a blinding light. That light would indeed blind us to the nature of the darkness. Rather, one goes into the darkness with a light that may be provided by the darkness itself—a death-light, shall we say, like the phosphorence of deep-sea creatures. Perhaps this death-light is none other than the torch of chaos and doubt we heard of earlier, taken up by the psychoanalyst as well in order to explore the divide that separates and links darkness and light, death and life. "I know that I have artificially blinded myself at my work in order to concentrate all the light on the one dark passage," wrote Freud to Lou-Andreas Salomé.[49]

We first know this death-light, in fact, in what Bion calls the mother's "reverie," her attempt to feel and make meaning of the baby's unspoken experience, including its wild distress and anxiety. The infant has no spoken language yet, and at times the reverie is less a peaceful daydream than an effort guided again by that torch of chaos and doubt, drawing on the mother's own knowledge of infant terror within the shelter of her own experience of survival and emotional connection. Making some kind of meaning of the baby's inchoate feelings, the mother restores a sense of order and safety in which life is larger than death. This capacity for emotional connection and a sense of continuity in relationship with another person allows us to acknowledge pain, loss, and death while still knowing that something extends beyond and encompasses these experiences.

"[Art] is always meditating on death," said Tomasi di Lampedusa.[50] The artistic force of wholeness can subsume and incorporate the darkness within the psyche. Deep beauty—in art, in the built environment, in the natural world—acknowledges death and the forces of destruction and brings them to our awareness in such a way that we become more able to tolerate and assimilate the pain of them. It is the successful work's capacity for integration that allows it to be a mirror of the self. This allows us to identify with the work enough to experience and learn from it and through it to be connected to the ground of all things. In the authentic work of art, the more the tragedy of life is included, the richer and deeper the work will be. The more integrated the work, the more it will be felt to harbor—but not to be dominated by—the most destructive forces the psyche knows: rage, terror, destructive greed and envy, cruelty, hatred, as well as the resulting guilt, depression, and deathfulness. "Art . . . bears witness to the world of depression or chaos overcome," says Adrian Stokes.[51]

As we mature, we become increasingly aware that excluding the forces of death and destructiveness somehow deprives the work of art of full participation in beauty. In a work where the forces of idealization run rampant, we sense that something has been forcibly excluded and that the work is thereby truncated and diminished. When we feel that a work is trivial, sentimental, or "merely pretty," we are sensing this superficiality and shallowness. Such works cannot touch our human core fully because they exclude the suffering that, as the Buddhists would say, is inseparable from human life.

In her futurist novel *Mara and Dann,* Doris Lessing provides us with a telling example of a work removed in this way from the processes of life. The exiled heroine of the book carries with her as a disguise a brown garment made of a synthetic material produced by a technology now lost. The garment does not wrinkle, decay, or even get dirty. A marvel of craft and wisdom? No—everywhere it evokes spontaneous feelings of discomfort and disgust on the part of those who see it and touch it. We need to see and acknowledge decay, the passing of time, birth and death. They are the emergence from nothingness and the passage back into it.[52]

In his paper "Death and the Mid-Life Crisis," Elliot Jaques links the issues of mortality, biological death, and the making of the successful work of art. What Jaques calls the midlife crisis has at its heart a confrontation with one's own mortality. Acknowledged and ultimately accepted, this awareness eventually brings about a serene resignation and detachment. In the realm of art, Jaques says, one finds in successful mature works a kind of serenity that reflects this acceptance.[53] This serenity contrasts with the fiery instantaneousness of more youthful works. Perhaps Dante's *Divine Comedy,* which depicts the poet's journey, at the midpoint of his life, through hell and purgatory to arrive in paradise, is the purest example of this kind of serene maturity in art. Jaques notes that the older, mature artist feels a continuity of self with his successful works that is not the case with the more throwaway productions of youthful fire. Perhaps the serenity that comes of an experienced acceptance of death and destruction is necessary in order to fully feel one's continuity with the nonhuman.

Ultimately, artistic form gains its beneficence from its capacity to integrate destructive and constructive forces. In doing so, it demonstrates that the wholeness and coherence of life, experienced as beauty, is stronger than destructiveness. It is this integration that provides hope and promises us strength as well as generosity from the experience of the work.

Teachings from the Void: The Artist, the Gift, and the Self

> To be an artist you need a teacher, and it is for this reason, above all, that I began collecting carpets.
> —Christopher Alexander, *A Foreshadowing of 21st Century Art*

If the ten thousand things emerge from the void, then a successful work of art, a work with the strange power to teach its own lesson, must also emerge from the void. Hence, Christopher Alexander claims, all beautiful works must be "elaborations of nothingness." If they are not elaborations of nothingness, they are jarring interpolations of something else into the natural unfolding of the void. The artist or maker, in other words, does not imitate or copy reality but rather participates in the process whereby things emerge from the void. That process must be embodied in what he or she makes. What does a human being have to do in order to work in a way that elaborates nothingness? How does the artist learn to do this?

The gift plays a decisive role in the process of learning. Teaching and gift giving are inseparable, for what the teacher gives the student is a gift, and what the student produces, if the process goes right, will be a gift given further, to the world. In common parlance, the artist is "gifted," passing the gift on in making art for the world. The gift that this kind of teaching represents is what Lewis Hyde would call a "transformative gift." It produces transformation in the one who receives it. In the case of the artist, the gift is often a work of art that itself incorporates and embodies the process. As Hyde reminds us, "Most artists are brought to their vocation when their own nascent gifts are awakened by the work of a master. That is to say, most artists are converted to art by art itself."[54]

The evolution of Alexander's capacity to "see" the early Turkish carpets he collects is the work of such a transformative gift. "I started collecting carpets, years ago, because of my desire, as a builder, to learn from them," writes Alexander in his book *A Foreshadowing of 21st Century Art.* "I felt that they had something to teach me, though I did not at first know what."[55] The carpets to which he was increasingly drawn seemed to succeed in the creation of a "being" that resonated deeply with his self and at the same time seemed to reveal the ground of being as such. As a builder, Alexander wanted to create things that would have the same effect. But the carpets were products of a tradition of craftsmanship with an ideal of beauty—a tradition nonexistent in the realm of building today. Lacking an analogous tradition, Alexander used those carpets as his teachers. This meant looking and looking at them and allowing his sensitivity to beauty to be refined and deepened by what he saw. As his sensitivity evolved, he traded his first carpets for others he perceived to have more to teach him. Thus his collection evolved with his capacity to be open to beauty.

The transformative action of the gift works through what Hyde calls "the labor of gratitude": "If the teaching begins to 'take,' the recipient feels gratitude. [Gratitude is] a labor undertaken by the soul to effect the transformation after the gift has been received."[56] In other words, the gratitude for the reception of the gift from the teacher is what motivates artists-to-be to make what they have received fully conscious and usable, to make it their own in such a way that they can give it farther. As we have seen, Hyde illustrates this with the story of the shoemaker and the elves. The elves' gift to the shoemaker and the shoemaker's efforts to understand the elves' capacity gradually result in the shoemaker's acquisition of the capacity to make beautiful shoes himself. At the point when he can give the gift of beautiful shoes, the teaching process has been completed.

Freud once spoke of "the impressive caesura of birth," and the phrase caught Wilfred Bion's attention.[57] There may be more continuity between the prenatal and the postnatal states than one might think, Freud had said. This and other caesuras may mislead us, Bion suggests. Just as we imagine ourselves unable to penetrate imaginatively to the other side of birth, we may imagine ourselves unable to penetrate other divides with our experience and our thinking. The transformative gift comes from across a divide that it helps us to penetrate. It comes from across the barrier that initially divides the artist-to-be from the void he or she must learn to elaborate. In the case of the shoemaker and the elves, it comes from across the divide between the spirit world and the human world, symbolized also by the divide between the night and the day consciousness. It is not until the shoemaker and his wife stay awake at night to watch the elves that the final step in the transformation occurs.

The gift from across the divide also reveals the underlying continuity between the two sides of the divide. There must, after all, be some kinship between the giver and the recipient for the gift even to be recognized as such. Remember that gifts create bonds. The act of giving the gift creates a bond that symbolizes and enhances this link with the other side of the divide. In this sense the gift is erotic, as in the subtitle of Hyde's book, *Imagination and the Erotic Life of Property:* "As gift exchange is an erotic commerce, joining self and other," Hyde writes, "so the gifted state is an erotic state: in it we are sensible of, and participate in, the underlying unity of things."[58] Participating in this way, we become aware of the emptiness from which form emerges. The link to the spiritual state is made through gifts, Hyde reminds us, as with gods who incarnate, offering their own bodies as gifts that establish the bond between human beings and the spiritual state in which the god participates.[59]

The makers of Alexander's carpets were Sufis concerned with making a gift to God that was also a revelation of God. In the carpets they made one perceives both a "mirror of the self" and "the luminous ground." The makers of the carpets, Alexander would say, received the gift of awareness of the spiritual ground. Seeing the self mirrored there, they could recognize their participation in the underlying unity of all things. Their gift to God brought something in return, both for them and for all those who learn from the carpets. The gift across the divide travels in both directions.

Meister Eckhart, the fourteenth-century Christian mystic, exemplifies Hyde's idea that through the labor of gratitude the spiritual gift yields union with the spiritual state. "For Eckhart," Hyde explains, "all things owe their being to God. God's initial gift to man is life itself, and those who feel gratitude for this gift reciprocate by abandoning attachment to worldly things, that is, by directing their lives back toward God." Those who empty themselves of the world receive a second gift: "a Child is born (or the Word is spoken) in the soul emptied of 'foreign images'."[60] This is the achievement of the labor of gratitude. "For Eckhart, the child born in the soul is itself a god," Hyde continues. "Whoever gratefully returns all that God has bestowed will, by that act of donation, enter the Godhead."[61] Ideally, Alexander says, agreeing with the Sufi makers of carpets, not only the mystic but also the maker of a work of art, recipient of the gift of artistic vision and capacity, is attempting to make a "gift to God," a gift in return to the source that has bestowed its gifts on him.

The Process of Unfolding as an Elaboration of Nothingness

> *The cosmic principles go back into silence, non-being, emptiness; a Nothing that will produce the ten thousand things, and the ten thousand things will have that marvelous emptiness still at the center.*
> —Gary Snyder, "The Great Clod Project"

We learn about beauty not only by studying beautiful things but also by attempting to make beautiful things. But it is not easy to make a gift to God. If we set out to make something beautiful we find that we can do so only by participating in a process of unfolding. "How can you make it more beautiful?" pianist Jeffrey Kahane asked his student. Given how the piece sounds now as you play it, that is, how and where could you make it still more beautiful—let it unfold still more into its potential beauty? While we may have an idea that guides us, if we are to make a living work, we cannot simply implement a fully preconceived idea; each step must be guided by what is already there. The work as it is at this point seems to call out for the next step in its unfolding, the next action that will make it more beautiful. By responding to its request, so to speak, we collaborate with the emerging work at each point as it unfolds. We allow the work to be an elaboration of the nothingness from which it is emerging. The process of unfolding is an elaboration, a labor of gratitude. We participate in the making, feeling our gratitude for the beauty of what is unfolding.

Attuning oneself to the process of unfolding means reorienting one's conception of the self. The beautiful work may be a "mirror of the self," but it is not a mirror of the limited self, the ego. This reorientation happens at the moment of what Joseph Campbell calls aesthetic arrest, the moment when the psyche receives the impact of the beautiful, "where all the faculties of sensation, thinking, feeling, and intuition are dissociated from the service of the artist's personal will, so that . . . he is released from fear and desire, because [he is] free (for the moment at least) of ego: 'beside himself,' 'transfixed by the object.'"[62] Making art sets our egotism in conflict with a different, larger, more participatory self that we might call our selflessness. The egotism that wants to set us apart and make us "original" and a "star" fights with the fact that the gifted state is participation in the underlying unity of things. The urge to be a star, to receive credit for originating the work, is antithetical to the labor of gratitude, which recognizes that the work that is a gift *to* God is in some sense the final realization of a gift *from* God.

Art, says Flannery O'Connor, involves a "very definite leaving behind of the niggardly part of the ego. . . . No art is sunk in the self, but rather, in art the self becomes self-forgetful in order to meet the demands of the thing seen and the thing being made."[63] In Meister Eckhart's terms, the attempt to be a star inserts "foreign images" into the labor. It inserts something extraneous—something coming from outside the gift exchange with God, so to speak—into the process whereby the work emerges from the void. Being original might seem to mean being authentically oneself, but in fact the focus on oneself is a distraction from the self-forgetfulness required by the actual process of receiving and laboring.

Self-forgetfulness of this kind means allowing one's self too to emerge from and pass into the void. The self too becomes a gift to be given on. For unlike commodities, gifts cannot be hoarded. One empties, as the mystics say, the self rather than hoarding it. As Rilke puts it, this entails "[n]ot any self-control or self-limitation for the sake of specific ends, but rather a carefree letting go of oneself; not caution, but rather a wise blindness; not working to acquire silent, slowly increasing possessions, but rather a continuous squandering of all perishable values."[64]

I am reminded again of the story about the Papago hunter and songs. In his talk "The Politics of Ethnopoetics," Gary Snyder poses the question, "How do you prepare your mind to be a singer"—in other words, to make a gift to God? "It takes an attitude of openness, inwardness, gratitude, plus meditation, fasting, a little suffering, some rupturing of the day-to-day ties with the social fabric," he says. Then he quotes from Ruth Underhill's book about the Papago (O'odham), *The Power of Song:* "The reward of heroism is not personal glory nor riches. The reward is dreams." The one who desired song did not focus on the song itself but rather on being pleasing to "the supernaturals" by being a good hunter or warrior. Then he might hear a song in his sleep: "Perhaps the clouds sing or the wind or the feathery red rain spider on its invisible rope."[65] The song, in other words, is conceived as a gift to the self rather than something created by the self. The gift does not bring the dreaming hunter personal glory—stardom. Rather, the hunter's receptivity, nourished through his attitude of openness, allows him to be intimately linked with both the natural world and his community. The clouds or the wind or the rain spider penetrate into his dreaming mind to produce their songs in him, and he gives his songs onward to be used in ceremonies. The bond between the natural world and the people is made by way of the conduit that is the selfless efforts and the open mind of the hunter.

There is an intimate connection between the selflessness I have been speaking of, and life as such. Priestley's dream showed us that. Neville Symington's conception of narcissism further illuminates the antithesis between the egotistical effort to be seen as a star and participation in the underlying unity of things. As we have seen, Symington thinks of narcissism as the result of rejecting the "Lifegiver," the source of life.[66] This rejection splits the self into a part that must, to survive, remain linked with the process of receiving and giving, and another part that has spurned the Lifegiver and thinks it can do and be alone. That part, however, being essentially cut off from life with its natural process of growth, development, and "increase," can only imitate others on a superficial and surface level. The deep source is cut off. The narcissistic part that has cut itself off from the life process, in other words, has only "foreign images" at its disposal. Paradoxically, in the maker of a work of art, the narcissistic will to be original, to be seen as a star in the eyes of others, can only result in the insertion of foreign images into the work.

As Bion pointed out, the preconceived idea, the idea that is already known, closes up the gap through which newness, the unknown, might emerge. The caesura of "as we all know" is a block to creativity.[67] When we let the work unfold step by step rather than serving our narcissism by imitating a preconceived idea in our heads, we allow something new and unknown to emerge. Following this process necessarily renders the work unique. We see this in the creation of the carpets in Alexander's collection. Why are the most beautiful also the rarest and most unique in design? Alexander asks. This is not accidental, he says, but the result of the maker's willingness to follow the unfolding process. It is extremely difficult for the artist to create a structure of such coherence that it has, in his words, "the power to 'raise' a being." "It can be done," he continues, "only by paying detailed attention to the uniqueness of a given situation." This intense and unique coherence tends not to arise when the artist is intent on copying a known pattern, but rather where some small bit of uniqueness in the work in progress takes off, as it were, and the carpet-maker follows as it leads in new and unexpected directions.[68] This kind of uniqueness, as we know, characterizes the natural world and its creatures—the "beings" it "raises." It is not only snowflakes that are never precisely the same. The natural diversity we tend to think of in terms of species extends down to the level of leaves and crystals and up to the level of landforms as well.

Personalness and presence are the companions of uniqueness. Being present oneself in the present moment to the work is essential both to the contemplation and to the making of beautiful things. This kind of presentness is the essence of personhood, the living individual. It may seem paradoxical that to participate without ego in the unfolding demanded of what is in the process of creation will create the sense of personalness and one's own personhood, but that is the case. In this sense too the work is a mirror of the self. It is as though life, livingness, recognizes personhood; as though life can only *be* in the form of personalness, and our own and that outside us reciprocally confirm one another.

Culture as Protector of Beauty

> *Satan's theft, preventing the lover from contemplating his Beloved.*
> —Henri Corbin

> *In the great period of carpet weaving, the apprenticeship of a dyer lasted fifteen years. At the end of this fifteen years, the apprentice dyer was required to make a color which no one had ever seen before. Only then would he, or she, be recognized as a master dyer.*
> —Christopher Alexander, *A Foreshadowing of 21st Century Art*

Aesthetic experience can be terrifying. What we call beauty has a force to it; it opens us up to the blazing furnace at the heart of the universe. That can feel like violence. Art invites us into a psychic world that provides not only beneficence and wholeness but also terror and destructiveness. Nature is life, but also decay and death. Hunting links us with the prey but at the cost of creating a dark hole in the day, as van der Post said. Selflessness and a mind open to the process of unfolding expose us to these intense forces and at the same time help us tolerate their impact. Human beings need sources of protection and strength to dare to move in the direction of openness. To some extent this protection and strength comes from inside, but initially and to some extent always, the person needs external support as well. The baby needs to be able to depend on the mother's capacity to tolerate his or her terrors. The traditions and rituals of indigenous societies need to help the adolescent tolerate the vision quest and the hunter to tolerate his role in causing death.

The human being who traffics with beauty also needs to be protected and nourished by culture and tradition. Meltzer speculates that the artworld with all its conventions, from formal openings to guards in the museums, helps to protect the viewer who responds to art's invitation.[69] We all know the importance of tradition to the artist, and the difficulty of achieving beauty without it. Cézanne depended on the Louvre and the Old Masters. The dyers who made the colors for the early Turkish carpets served a fifteen-year apprenticeship. Where culture fails, beauty is endangered. As Alexander tells us, it is now nearly impossible to make a beautiful building; the traditions that would support that goal no longer exist. People have given up on beauty as the aim of their work in architecture.[70]

Of course, there is always a tension between cultural forms, which necessarily imply some rigidity, and the emergence of the new. The new can never be completely contained by the old and established traditions. Thus the new always raises questions and puts the old in doubt.[71] Where culture and tradition do shelter the possibility of encountering and making beauty, it is not by imposing rigid rules. Rather, they provide an extra layer of protection for the psyche that wants to open itself to larger forces. In doing so, they make it possible for that psyche to join with the process of creative unfolding that will allow something new to emerge. Just as babies need a safe environment and the containing mind of the mother in order to discover their own selves in their capacity to be alone (first in the presence of another), so it is very difficult to come to beauty, to have a guiding conception of what one is aiming for, without the "container" of a tradition and a community. Culture, in other words, can both protect the psyche from inhibitions *against* exploring and give the explorer a sense of aspiration. "It is not possible to be original except on a basis of tradition," Winnicott insists; "conversely, no one in the line of cultural contribution repeats except as a deliberate quotation, and the unforgivable sin in the cultural field is plagia-rism."[72] To complete his fifteen-year apprenticeship, remember, the dyer had to create a color that had never existed before.

Culture itself is a kind of second-order aesthetic experience. "Just as the cre-ation of art is a love relationship with the world, the creation of culture and so-ciety is a loving mastery of the ambivalence of self and not-self," writes Shepard.[73] Culture is able to protect the experience of self-in-relationship that is central to aesthetic experience precisely because it is essentially an elabora-tion of the baby's earliest experience of separateness-in-relationship—the experi-ence of play. Babies discover play when they become able to link separateness and union in their relationship with the mother. Where in the mind is play? asks Winnicott in his essay "The Location of Cultural Experience." He defines its location as the point where the object is used as a symbol of the relation-ship between separateness and union: "The use of an object symbolizes the union of two now separate things, baby and mother, at the point in time and space of the initiation of their state of separateness."[74]

"Potential space" and "the third area" are names Winnicott gives to this area in which play occurs. This is the locus of creativity. Experience in this space has the quality of unfolding that characterizes the process through which wholeness emerges. For the baby, in this space on the border between merging and separateness, the new can emerge and still be experienced as the baby's own, created in a context of trust and safety. In this space of play, everything feels new and creative to the baby, just as it feels intense and personal:

> The baby finds intense, even agonizing, pleasure associated with imaginative play. There is no set game, so everything is creative, and although playing is part of object-relating, whatever happens is personal to the baby. Everything physical is imaginatively elaborated, is invested with a first-time-ever qual-ity. . . . For the baby (if the mother can supply the right conditions) every detail of the baby's life is an example of creative living. Every object is a "found" object. Given the chance, the baby begins to live creatively, and to use actual objects to be creative into and with.[75]

Cultural experience, Winnicott says, is a later elaboration of this area, provid-ing a bridge between an internal world of personal subjectivity and an external world that is quite separate from us.

Play may well be the area of experience that most puts us in touch with what life is all about, writes Winnicott. Indeed, the third area may be the area of experience that puts us in touch not only with human life but with all life, allowing us to not only understand and appreciate it but to experience a personal kinship with it. For play goes beyond instinct and appetite to allow us to participate in larger dimensions of life and selfhood. Play, and the cultural experience that develops from it, can be intensely pleasurable—the body and the mind are both engaged—without being involved in satisfying an appetite or an instinctual desire. Based on the experience of separateness within union, play allows us to feel relatedness with the other, whatever form it may take, rather than simply using the object for our own needs on the one hand or observing it from the outside on the other.

But of course cultures, and individual traditions and rituals as well, can fail, grossly or in small ways, just as the mother's caring can fail the child. To be able to play, the baby must have confidence in the mother. He or she must trust in the continuity of experience with her. When the mother leaves, the baby's internal image of her fades after a time and the baby becomes distressed. When the mother returns soon enough, her image in the baby's mind is restored, and the distress is relieved. In this way separation becomes tolerable and even useful. The baby builds up a capacity to play and to return to playing. But when the mother stays away beyond the point where the baby can tolerate the pain of her absence, the baby becomes traumatized and loses the sense of continuity.

Without confidence that the mother will reliably enough return before her absence becomes traumatic, the baby becomes unable to play. When this happens, the space the baby might have used for creative play becomes vulnerable to invasion by the contents of someone else's imagination—invasions that are experienced as alien and persecuting, but that the baby feels helpless to prevent. These are the "foreign images" Meister Eckhart warned against—other people's dreams, as Thomas Berry calls them. A baby whose mind is filled with such foreign implants easily becomes an adult who is unable to contribute to the potential space that is culture and to participate in the unfolding process that puts us in touch with life. A society shaped by such adults would be incapable of aiming at beauty. In such a society, the cultural space itself might be usurped by the "dreams of others," shaped by the violence of trauma and deprived of its true function. I believe this is in fact to a large extent our situation.

Can Culture Constrain Evil?

Our presence has become so deadly.
—Thomas Berry

The question of evil, like the question of ugliness, refers primarily to the anaesthetized heart, the heart that has no reaction to what it faces, thereby turning the variegated sensuous face of the world into monotony, sameness, oneness.
—James Hillman, *The Thought of the Heart and the Soul of the World*

Culture, and aesthetic experience in general, carries the capacity for containment and meaning making originally created in the aesthetic reciprocity between baby and mother. It too is located at the border between connectedness and separateness where pain and destructiveness can be felt and the new can unfold in some degree of safety. As the communal locus of the concern for life and truth, culture can give containing form to the violence that characterizes instinctual experience and the emotions of love and hate. As Christopher Bollas points out in his discussion of Sophocles's *Oedipus* cycle, cultural forms can contain psychic complexity of a degree that cannot be mastered by the individual psyche, thus making it possible for individuals to "play" with issues that would otherwise overwhelm them with anxiety.[76]

There are dimensions of mental violence that attack the very possibility of separateness-in-relatedness that culture and the aesthetic dimension represent. To cut the link that exists by nature, to spurn the Lifegiver, as Symington says, is an act of this kind of violence. The rejection of connectedness can take the form of a literal discarding or evacuation of what is within us. In psychoanalytic terminology, this is called *projection*. C. Fred Alford considers projection the essential mechanism in evil. Those who have been put in a state of dread—and by dread he means something worse than fear, a dread that eradicates all sense of foundations—can decide to project it outward into others, in short, to evoke dread in other people.[77] Inspiring dread and causing death are intimately related, though not identical. Evil is the inspiration of deathfulness in others. The deathful darkness that van der Post and Raine described was an experience tinged with dread.

This deliberate refusal to acknowledge connectedness, this desire to project utter separateness, may arise as revenge for some frustration or trauma. Perhaps our presence has become so deadly because we are taking revenge on the natural world, as though it were a mother we want to spurn. Perhaps in "our inner rage against the conditions of our earthly existence," as Thomas Berry puts it, we want to traumatize the natural world as we ourselves feel we have` been traumatized.

Evil is the product of the anesthetized heart, writes Hillman.[78] The aesthetic heart is the heart that responds to the beguiling face of nature with an imaginative response. This is the creative looking Winnicott was talking about, and the reciprocal love making with nature that intensifies the sense of kinship. The anesthetized heart, in contrast, is cold, and cuts off its response, spurning the Lifegiver. This cold eye, this anesthetized heart—are they not on their way to wanting to inspire dread in what they gaze on?

Just as a mother can take a baby's primitive terror—the fear of dying, of being annihilated, of falling into a black hole or an abyss—and contain it so that it becomes bearable for the baby as fear but not infinite abyss-like terror, so it is possible, Alford argues, for a culture to contain the experiences of dread that enter into evil. Greek tragedy provided such a cultural container, he believes. We can think of art, he suggests, as a way to give form to our dread. This "is what the ancient Greeks . . . did, confronting the sheer terror of existence, their dread, with nothing but beauty, the beauty of their art and the graceful beauty of their lives."[79]

We may, however, now be living in a culture that does not have adequate arenas within which to contain and neutralize our contemporary traumas and dreads.[80] If that is so, it may be our dreads and our deathfulness rather than our love making that escalate and reverberate as the natural world we destroy reflects back to us our loss of self. In ancient Greek culture, Michael Ortiz Hill reminds us, those who wounded Mother Earth would be pursued by the avenging Furies. "Perhaps now the Furies are gathering around us again in the form of multifarious ecological catastrophe," Hill writes. "Meeting the accusing eyes of the Furies," he adds, "could be the first harsh, but necessary, lesson in learning to live peaceably with [the] soul [of the world.] Nature then would not only be something we 'look at' but also a presence that observes us and, by surrounding us, teaches us the true dimensions of an ecological niche within which our species can thrive."[81]

Notes

1. Donald Meltzer, *Studies in Extended Meta-Psychology*, 204. Stephen G. Shehorn suggests that the baby, and the mother when she was a baby, began with a "primary identification with the boundless infinite," which enters into the aesthetic reciprocity they experience together: "Before there is a 'personal' self or a personal mother, there is an ocean of experience potential of terror/bliss. . . . Depending on how the mother is herself established in that primary background energy so that it shines through her countenance, the infant's experience of terror/bliss [does or does not become] bearable" ("The Problem of the Self in Psychology," 41).

2. Meltzer and Williams, *The Apprehension of Beauty*, 22.

3. Meltzer and Williams, *The Apprehension of Beauty*, 22.

4. See Joseph Campbell, *Creative Mythology* (vol. 4 of *The Masks of God*), 659.

5. Meltzer and Williams, *The Apprehension of Beauty*, 6.

6. Gary Snyder suggests the role of aesthetic reciprocity in the hunt, for instance. "Man is a beautiful animal," he tells us. "We know this because other animals admire us and love us. Almost all animals are beautiful and paleolithic hunters were deeply moved by it. . . . Hunting magic is designed to bring the game to you—the creature who has heard your song, witnessed your sincerity, and out of compassion comes within your range" ("Poetry and the Primitive," in *Earth House Hold,* 120).

7. Meltzer, *Studies in Extended Meta-Psychology,* 208.

8. See Meltzer and Williams, *The Apprehension of Beauty,* chap. 4, "On Aesthetic Reciprocity."

9. Michael Eigen, "The Significance of the Face," in Adam Phillips, ed., *The Electrified Tightrope,* 59.

10. D. W. Winnicott, "Mirror-role of Mother and Family in Child Development," in *Playing and Reality,* 111–118.

11. Winnicott, "Mirror-role of Mother and Family in Child Development," in *Playing and Reality,* 114.

12. Winnicott, "Mirror-role of Mother and Family in Child Development," in *Playing and Reality,* 115.

13. Paul Shepard, interview in Derrick Jensen, *Listening to the Land,* 250.

14. For a brief exposition of this idea, see Christopher Alexander, *A Foreshadowing of 21st Century Art,* 21–30. For a fuller exposition see Alexander's forthcoming *The Nature of Order.*

15. Alexander, *A Foreshadowing of 21st Century Art,* 22.

16. Adrian Stokes, "Form in Art," in Melanie Klein, Paula Heimann, and R. E. Money-Kyrle, eds., *New Directions in Psychoanalysis,* 407.

17. See for instance, Stokes, "Form in Art," in Klein, Heimann, and Money-Kyrle, eds., *New Directions in Psychoanalysis,* 406–420, and "The Invitation in Art," in *Critical Writings,* vol. 3, 267, 271. "Form in Art" in particular expounds the benignity of aesthetic experience.

18. See Stokes, "The Invitation in Art," in *Critical Writings,* vol. 3, esp. 270–272.

19. Stokes, "The Invitation in Art," in *Critical Writings,* vol. 3, esp. 270–272.

20. Meltzer and Williams, *The Apprehension of Beauty,* 192. The whole of the chapter from which the quotation is taken, "Holding the Dream," is relevant here.

21. James Hillman, *The Thought of the Heart and the Soul of the World,* 45.

22. Hillman, *The Thought of the Heart and the Soul of the World,* 47.

23. Hillman, *The Thought of the Heart and the Soul of the World,* 48.

24. Quoted in Walter Benjamin, "On Some Motifs in Baudelaire," in *Illuminations,* 188.

25. Hillman, *The Thought of the Heart and the Soul of the World,* 115.

26. For a treatment of the notion of the *mundus imaginalis,* or imaginal world, in Persian mysticism, see Henry Corbin, *Spiritual Body and Celestial Earth,* esp. vii–xxix, 3–105.

27. Corbin, *Spiritual Body and Celestial Earth,* 11.

28. Corbin, *Spiritual Body and Celestial Earth,* xiii.

29. Benjamin, "On Some Motifs in Baudelaire," in *Illuminations,* 188.

30. Quoted in Benjamin, "On Some Motifs in Baudelaire," in *Illuminations,* 188–189.

31. Benjamin, "The Image of Proust," in *Illuminations,* 204.

32. Jack Turner, *The Abstract Wild,* 15.

33. Snyder, "The Great Clod Project," in the *Gary Snyder Reader,* 293–294.

34. Quoted in Edward Edinger, *Anatomy of the Psyche,* 129–130.

35. Quoted in Shehorn, "The Embodied Countertransference," 58.

36. Charles Rosen, *The Classical Style,* 448.

37. Rosen, *The Classical Style,* 446–447.

38. Quoted in Snyder, "Blue Mountains Constantly Walking," in *The Practice of the Wild,* 104–105.

39. See, for instance, John Seed et al., *Thinking Like a Mountain.*

40. See Freud's remark in *The Interpretation of Dreams,* quoted in James Hillman, *The Dream and the Underworld,* 140: "There is at least one spot in every dream at which it is unplumbable—a navel, as it were, that is its point of contact with the unknown."

41. Snyder, "Blue Mountains Constantly Walking," in *The Practice of the Wild,* 104–105.

42. Snyder, "Blue Mountains Constantly Walking," in *The Practice of the Wild,* 104–105.

43. See Corbin, *The Man of Light in Iranian Sufism,* chap. 5, "The Black Light."

44. Alexander, *A Foreshadowing of 21st Century Art,* 22.

45. Snyder, "Blue Mountains Constantly Walking," in *The Practice of the Wild,* 110.

46. Laurens van der Post, *Yet Being Someone Other,* 69–70.

47. Kathleen Raine, *Farewell Happy Fields,* 61.

48. Raine, *Farewell Happy Fields,* 61–62.

49. Freud, letter to Lou Andreas Salomé of May 25, 1916, cited in Wilfred Bion, "Notes on Memory and Desire," in *Cogitations,* 384.

50. Giuseppe Tomasi di Lampedusa, *The Leopard,* 18–19.

51. Stokes, "Form in Art," in Klein, Heimann, and Money-Kryle, eds., *New Directions in Psychoanalysis,* 413.

52. Perhaps architecture is felt to be the most fundamental of the arts because the built environment shows us decay all around; we live within it. The inherent limitations of humans' capacity to build, the need to build things human scale, and the stresses and strains and decay processes built into the structure of matter itself all demonstrate impermanence to us. And perhaps so much of the modern built environment is experienced as dehumanizing because it tries to conceal or deny these processes through monumentality of scale, synthetic materials or materials intended to look as though they cannot decay, and structures that seem to defy or deny gravity.

53. Elliot Jaques, "Death and the Mid-Life Crisis," in Elizabeth Bott Spillius, ed., *Melanie Klein Today,* vol. 1, 246–247.

54. Lewis Hyde, *The Gift,* 47.

55. Alexander, *A Foreshadowing of 21st Century Art,* 17.

56. Hyde, *The Gift,* 47.

57. See Bion, "On a Quotation from Freud," in *Clinical Seminars and Other Works.* The quotation from Freud reads, "There is much more continuity between intra-uterine life and earliest infancy than the impressive caesura of birth would have us believe." See also Bion's essay "Emotional Turbulence," in the same volume.

58. Hyde, *The Gift,* 163.

59. Hyde, *The Gift,* 58.

60. Hyde, *The Gift,* 54.

61. Hyde, *The Gift,* 55.

62. Campbell, *Creative Mythology* (vol. 4 of *The Masks of God*), 659.

63. Quoted in Hyde, *The Gift,* 150.

64. Quoted in Hyde, *The Gift,* 150.

65. Ruth Underhill, *The Power of Song,* quoted in Gary Snyder, "The Politics of Ethnopoetics," in *A Place in Space,* 140.

66. Again, this is the fundamental idea of Neville Symington's *Narcissism.*

67. See Bion, "Emotional Turbulence," in *Clinical Seminars and Other Works,* 303.

68. Alexander, *A Foreshadowing of 21st Century Art,* 84.

69. See "Concerning the Social Basis of Art," a dialogue between Adrian Stokes and Donald Meltzer, in Stokes, *Painting and the Inner World,* 35.

70. Alexander, A *Foreshadowing of 21st Century Art,* 18.

71. Bion elaborates on this tension in terms of the tension between "the mystic" and "the establishment." Ideally, the establishment finds a way to contain and make good use of the mystic's new idea, but the relationship between the two is also potentially destructive. The establishment can crush or suffocate the new idea, and the new idea can explode and obliterate the establishment. See Bion, *Attention and Interpretation,* esp. 72–82.

72. Winnicott, "The Location of Cultural Experience," in *Playing and Reality,* 99.

73. Paul Shepard, *The Tender Carnivore and the Sacred Game,* 169.

74. Winnicott, "The Location of Cultural Experience," in *Playing and Reality,* 96–97.

75. Winnicott, "The Location of Cultural Experience," in *Playing and Reality,* 101.

76. See "Why Oedipus?", in Christopher Bollas, *Being a Character,* 218–246.

77. This is the argument of C. Fred Alford's *What Evil Means to Us.*

78. Hillman, *The Thought of the Heart and the Soul of the World,* 66.

79. Alford, *What Evil Means to Us,* 129.

80. I do not mean to imply, of course, that this would be the first time in history that culture has failed in this way.

81. Michael Ortiz Hill, *Dreaming the End of the World,* 72.

5   A Severe and Pervasive Apathy: Trauma, Destructiveness, and the End of the World

Even beyond the threat of nuclear warfare, I think, the ecological crisis is the greatest threat mankind collectively has ever faced. . . . My hypothesis is that man is hampered in his meeting of this environmental crisis by a severe and pervasive apathy which is based largely upon feelings and attitudes of which he is unconscious.

—Harold Searles, "Unconscious Processes in the Environmental Crisis"

In his book *Dreaming the End of the World,* Michael Ortiz Hill points out that nuclear threat and environmental destruction are intertwined in the dreams he collected: "The psyche has interpreted the presence of the Bomb as a specifically ecological catastrophe."[1] Indeed, in trying to understand the emotional impact of the pervasive environmental destruction we live with, I have learned the most from analyses of the psychological effects of what I will call large-scale social catastrophes—the Holocaust, atomic bombings and the nuclear threat, and war, as well as individual experiences of severe trauma and abuse. This continues to surprise me. Apparently it is difficult for me to accept that the impact of environmental degradation is so far reaching and so devastating. Indeed, few of us are accustomed to thinking of it in these terms. Of course there are great and significant differences among these catastrophic experiences. But the intertwining of guilt and victimization, the feeling of being somehow tainted by death, the sense of something fundamental to ongoing life having been ruined, the feeling of being isolated and silenced, the wish to shut off the emotional pain on the one hand and the urgency to reintegrate and make meaning on the other—all these dimensions of response run through these catastrophic experiences and their aftermath and continuation. At some level, perhaps still more muted, I believe, they also characterize our responses to environmental degradation.

The End of the World as the Loss of Symbolic Immortality

> *[The survivor] mourns also for inanimate objects and lost symbols, for possessions, houses, streets he had known, beliefs that have been shattered, a way of life that has been "killed."*
> —Robert Jay Lifton, *Death in Life*

Robert Jay Lifton argues that human beings cope with the fact of mortality by finding avenues for what he calls "symbolic immortality"—ways they identify themselves with larger dimensions that are not subject to death as the individual human being is. These domains confer on life a meaning that transcends the individual's own empirical life. Among those avenues, Lifton lists one's descendants, the afterlife guaranteed by one's religious beliefs, cultural achievements, and the natural world, all of which survive the individual human being and absorb the individual in some transcendent process or domain.[2] The security provided by these forms of symbolic immortality makes one's individual mortality bearable.

But as Lifton points out, none of those domains of symbolic immortality any longer seems secure in its own prospects for continuation. All are eroded or under threat. We are all aware of the fraying of the mainstream religions. The nuclear threat and now the environmental threat have cast the continuity of human generations and the natural world, and with it the cultural world, into question. The poet Rainer Maria Rilke, writing in 1919, after the end of World War I, describes this lost security in terms of the sheltering constellations of the sky. "I still do not know and have not been able (for how long) to give myself the slightest proof of inner activity," he writes. "The intersecting point of my forces has lost its starriness, has fallen out of the great constellations that used to shelter and support it in spiritual space."[3] Psychologically, then, "the end of the world" means the erosion or destruction of the domains of existence that offer the hope of symbolic immortality. Rather than a unified world structure that supports us in spiritual space, to use Rilke's words, there is a unified sense of threat to our world.

In his discussion of the psychological impact of nuclearism, by which he means our ongoing involvement in nuclear weaponry and the possibility of the use of nuclear weapons, Lifton distinguishes between what he calls the "apocalyptic" self and the "measured" self.[4] The measured self is the everyday familiar self, concerned with the individual's life in ordinary time. The apocalyptic self is constituted by and concerned with the threat to life as such. It is both terrified and terrifying. Hill speaks to the same phenomenon: "The psyche is sheltered and contained by culture, land, kinship network, home—a background of social reality from which it gains coherent meaning," he writes. "There is no way to exaggerate the stark, raw nakedness of the apocalyptic psyche when it has lost this encompassing 'home.'"[5] Just as all contemporary people are living with both these selves, so they are living in two worlds, two realities. "Just as the Australian aborigines speak of a 'dreamtime' parallel to ordinary, mundane reality, I take the geography of apocalypse to be a real and vivid territory running alongside or beneath the day-to-dayness of our lives," writes Hill. "We live, in effect, in two worlds—our daily preoccupations and activities; and behind that, a fantastic landscape of terror or ecstasy: invisible poison, mutant fish, random violence or poignant tenderness; underground shelters, blistered skin, lost children; ashes, ruination, or grave mystery."[6] Or, as Arthur Colman more succinctly puts it, "Reality in the nuclear era is simultaneously as it has been and profoundly different from before."[7] The world as experienced by contemporary Europeans and North Americans, certainly, will never be the same. At some level of awareness, a sense, however illusory, of the security and "ordinariness" provided by confidence in the continuity of life has vanished.

In a situation in which life as such is in danger, both the measured and the apocalyptic selves are problematic. The adaptive processes that serve to protect the measured self give rise to destructiveness, while direct responses to the apocalyptic threat can endanger the continuation of the everyday self. Yet we need both selves. We need the ordinary self to remember that life in its simple living is of value, and we need the apocalyptic self for its awareness of potential catastrophe. As Lifton says, "Hiroshima was an 'end of the world'. . . and yet the world still exists. Precisely in this end-of-the-world quality lies both its threat and its potential wisdom."[8]

Unraveling the Fabric of the World: Trauma and Betrayal

> *In the nuclear era, the archetype of Death-and-Rebirth is being replaced by*
> *what I term Death-without-Rebirth.*
> —Arthur Colman, *Up from Scapegoating*

In trauma we get a direct view of the "end of the world" in this sense. For in a very immediate way trauma destroys the individual's sense of a safe world in which to live. It is as though the individual is enabled to live life through being "held" within a containing context that is both cultural and natural— a "world." This world is in the background, and provides a foundation and a shelter, as it were. One might think of it as a background fabric that holds the individual, a fabric woven of various contextual threads, from family to community to natural world to divinity, all imbued with a sense of strength, goodness, and authority, all permeated with the energy of life.

The most basic feature of trauma is the destruction of some part of this fabric through a betrayal of trust in something beyond the self.[9] In the betrayal there is a sense of the destruction of a broader authority, morality, or community. Trauma touches levels from the family to the community to the society to humankind to the natural world to the divine. Evil enters into every trauma, for what is betrayed is ultimately our trust in goodness, or the capacity of goodness to withstand evil, the strength of life as such. In the sexual abuse of children what is betrayed is trust in parental protection. In war what is betrayed is the sense of a country that one can believe in and that has one's interests at heart—parental protection in adult form. In the Holocaust it was the betrayal of confidence in human beings' limits on evil, or confidence in a good God who could and would protect the world from absolute evil. In the nuclear threat what is at stake is the betrayal of the sense that the world as such will go on, that there will be a world to be part of. In environmental degradation we experience the betrayal of the sense that life as such will continue, a betrayal of our confidence in the strength of the containing natural world that is the mother, so to speak, and life support of us all.

The betrayal experienced in trauma takes place against a background of dependency accompanied by a heightened sense of vulnerability. It is this combination of extreme dependency and extreme vulnerability that makes the betrayal in trauma so catastrophic. Think how dependent abused children are on their parents, and how vulnerable to their brutality. Or how much the soldier, if he or she is to have any chance to survive, depends on the larger military organization to supply weapons that function and orders that are rational. Bruno Bettelheim calls such situations, where one is extremely vulnerable and dependent and at the same time completely powerless either to ensure one's one survival or to help others, "extreme situations." Extrapolating from his experiences in a Nazi concentration camp, Bettelheim lists these characteristics of the extreme situation: "its shattering impact on the individual, for which he was totally unprepared; its inescapability; the expectation that the situation would last for an undetermined period, potentially a lifetime; the fact that, throughout its entirety, one's very life would be in jeopardy at every moment; and the fact that one was totally powerless to protect oneself."[10]

We tend to think of trauma as a single event out of the ordinary that is experienced as a violent assault on ordinary life. But all of us now live with environmental degradation all the time. In one sense, then, this assault on life is not something out of the ordinary. Environmental destruction (and remember that for the psyche, the nuclear threat and ecological catastrophe are intertwined) implicitly represents not only individual death on a massive scale but also an end to the processes of life, an "end to birth," as extinction has been called. Environmental degradation, and the nuclear threat that is intertwined with it in our minds, represent a trauma of a higher order, one that shares with the extreme situations Bettelheim described the expectation that it will last an indeterminate period, potentially a lifetime, a trauma that raises the possibility of the end of ordinary life as such and signals the advent of an ongoing potential "end of the world."

The Interplay of Inner and Outer Reality

The "world-destruction" fantasies of the psychotic reflect his radically impaired relationship to the world, and his projection upon it of his own inner sense of "psychic death." But the survivor of mass death reverses the process so that an overwhelming external *experience of near-absolute annihilation makes contact with related tendencies of the inner life.*
—Robert Jay Lifton, *Death in Life*

This "outer" reality is psychologically as much a part of us as its poisonous waste products are a part of our physical selves.
—Harold Searles, "Unconscious Processes in the Environmental Crisis"

At the same time he stresses the magnitude of the environmental crisis—the greatest threat mankind has ever faced, he calls it—Harold Searles points out how devastatingly inner factors work together with that threat. "My hypothesis is that man is hampered in his meeting of this environmental crisis by a severe and pervasive apathy which is based largely upon feelings and attitudes of which he is unconscious,"[11] he writes. It is not only the objective magnitude of the threat that makes it so deadly; it is the resonance between the psychological impact of the threat and internal forces of which we are not aware.

Searles points here to a complex interplay between internal factors and external reality, and to the way that interplay can keep us involved in being both victims and perpetrators of environmental destruction. It is obvious, he would say, that there is a severe external problem of environmental degradation. All our ways of examining external reality tell us so, though we may want to debate matters of degree and prediction. It is equally obvious that we are apathetic about this severe external problem. We go on living in such a way that the problem becomes worse, and we do not put it at the forefront of our consciousness and our concern. This response is highly irrational in terms of our welfare and survival. Such an irrational response signals an internal problem: something is interfering with our ability to deal with an important threat.[12]

What interferes, to follow Searles's argument further, is that rather than deal with our internal emotional conflicts, we project them outward and see them in the external world. Because we cannot in fact resolve internal matters by way of the external world, however, but only in essence reenact the internal conflict through the external "players," the actual external problems are neglected. It is as though we simply use the external world to enact and reenact our emotional conflicts, at whatever cost to the real world.

Jacques Lusseyran tells a story that illustrates this idea of projecting the internal world onto external reality. In an essay on his experiences in Buchenwald, Lusseyran, a leader of the French Resistance (and incidentally, blind from the age of eight and an adolescent during his service in the Resistance), tells about his encounter with an elderly welder from the Jura region of France named Jeremy. Jeremy states his understanding of the concretization of inner states in the outer world in the starkest of terms: Jeremy, Lusseyran reports, "had always seen people living in fear and in the most invincible of all fears: that which has no object. He had seen them all desire secretly and above all else one thing, to do harm to themselves. It was always, it was here, the same spectacle. Simply, the conditions had finally been completely fulfilled. The war, Nazism, the political and national follies had created a masterpiece, a perfect sickness and misery: a concentration camp."[13]

Among the examples Searles himself provides of unconscious feelings played out in the external world is the older generation's feelings of envy for the younger generation—envy of people who have, at least in the fantasy of the elders, greater strength and health and more life ahead of them. Sending the younger generation off to war, or letting the world be spoiled at a rate that will make greater difficulties for the next generation than for the current one, are ways of taking revenge—in fantasy, mind you—on the younger generation, ways of spoiling the wonderful life that the older generation imagines and envies.[14] Consciously, of course, the older generation cares about the younger. This caring is real, and in its rational moments the older generation would certainly not wish to spoil the world for the younger. But envy that is not acknowledged and dealt with as a feeling can be acted out in this way. Some of the poetry that emerged from World War I bears witness to this, and Generation X too is quite aware that their elders have bequeathed to them a damaged world.

At the same time as we use the external world to enact our internal emotional conflicts, Searles says, the actual degradation of the external world exacerbates those conflicts. It arouses fear, guilt, and dread of an overwhelming death. But in a vicious cycle, the more we are terrified by the state of the external world, the less we retain the ability to thoughtfully separate our internal conflicts and our external situation and deal with each in a manner appropriate to it. Both the nuclear threat and environmental degradation arouse a fundamental catastrophic anxiety, evoking the infant's fear of annihilation. These very early anxieties too are projected outward. Or as Lifton explains it with regard to the Hiroshima experience, the external experience of near-absolute annihilation "merges with mental images which originally signified the 'end of the world' for the young child—threatening images not only of separation and helplessness, but also of stasis and annihilation."[15] Thus the habit of unconsciously projecting our internal conflicts onto the external world continues. It is only by discriminating between the inner and the outer, and by reowning, so to speak, our projected internal conflicts and dealing with them thoughtfully, that we can allow ourselves to attend to the objective degradation of the external world.

Identifying with the Machine: The Fantasy of Omnipotence

The proliferation of technology, with its marvelously complex integration and its seemingly omnipotent dominion over nature, provides us with an increasingly alluring object upon which to project our 'nonhuman' unconscious strivings for omnipotence.
—Harold Searles, "Unconscious Processes in the Environmental Crisis"

One of the most important conflicts we project onto the outer world, Searles would say, is the conflict between our separateness and humanness on the one hand and our continuity with the nonhuman in us on the other. To avoid destroying the outer world, we need to become aware of that internal conflict, and to learn to relate these two aspects of ourselves to one another.

At the beginning of life, Searles says, we do not differentiate between the human and the nonhuman, the organic and the inorganic. That is in fact the truth of our physical being: our bodies are both organic and inorganic, in continuity with the rest of the material world.[16] As we grow, develop, and become separate human beings, we need to separate ourselves subjectively from the nonhuman. But this process of separation is a complex and painful one. It involves an awareness of our own limitations and mortality. It involves being in a state of permanent uncertainty and needing to make choices. We deny our relatedness to the nonhuman because we fear sinking back into lack of differentiation if we remember our continuity with it. Nothing is as dreaded by a human being as feeling like an inanimate object. That would mean having no feelings, being cut off from human life. But at the same time we long to ease the pain of separateness by becoming nonhuman again. Hence we are tempted to identify with the fantasied omnipotence of the machine.

How uneasy we are with this ambivalence! Our complex relationship to the nonhuman, exemplified in our relationship to the machine, lures us to destructiveness, Searles says. No individual can "save the earth." In trying to work against environmental destruction each of us is just one small person working with countless others. The fantasy of omnipotence speaks to our wish to escape such protracted and arduous efforts. Destruction offers itself to our fantasy as a form of omnipotence. We imagine it to be effortless and instantaneous. This is why the bomb is the perfect symbol of it. When we imagine ourselves presented with a choice between effortful attempts at restoration and understanding on the one hand, and an instant end to suffering through destruction on the other, it is tempting, Searles points out, "to give oneself over to secret fantasies of omnipotent destructiveness, in identification with the forces that threaten to destroy the world."[17]

The Future Foreclosed and Death Denied

> *It is only where tormenting and unassimilated experience blocks the develop-*
> *ment of the mental life that the vastness* of the human concern is stifled,
> and the mind is poisoned or goes dead.
> —Stephen Shehorn, "The Problem of the Self in Psychology"

> *A usurping of psychic actuality by the original experience [of death immersion]*
> *to such a degree that past and present are confused. . . .*
> —Robert Jay Lifton, *Death in Life*

When trauma or catastrophe destroys our sense of a safe, good, and strong world that can hear and bear our pain, we ourselves become unable to "suffer" that pain. We cannot experience it, digest it, and contain it within our own larger selves. And if we have no way to suffer our pain, we cannot mourn the dead. Without mourning, we have no way to move on within a world that remains meaningful. For death to become real and mourning to proceed, the individual must be able to attend to the individual death and the feelings it evokes. In a large-scale catastrophe, however, the person may be so utterly immersed in death that this is not possible. The dead become a mass, a heap, a cloud. As Lifton puts it, the survivor senses that "the bodies . . . around which he might normally organize rituals of mourning [have] abruptly disappeared into smoke or nothing."[18]

Time brings death; to live in time is to be mortal. To mourn is to complete the action of life, death, and loss. The loss of a sense of time means an inability to mourn, and the loss of a sense of time is a major effect of trauma. An action carried to meaningful completion is filled with, and takes place within, a framework of ongoing, substantive time. A traumatic experience has no such qualities; a barely endurable state is experienced as timeless, never ending. Traumatic experience, therefore, stands outside the normal course of time. With the loss or rejection of time we obliterate the future as well. The amorphousness of blank catastrophe, the loss of shape and structure in the social world, renders the sense of moving toward a future meaningless. Since the experience cannot be completed, there can be no sense of a future. Thich Nhat Hanh was right to call one of his books *For a Future to Be Possible.*

Think how important the ability to complete an action is to the small child developing the sense that he or she lives in time. Winnicott gives us a glimpse of this when he describes how a ten-month-old baby becomes interested in a spoon, makes it his own by taking it in his mouth and seeing what he can do with it, moves to a phase of dropping it, and finally loses interest altogether.[19] Something has proceeded through its course, from beginning through middle to end. The end is indeed a loss, a small death—the loss of interest in the spoon. But that is the natural end of the original interest, and it leaves room for the next interest to arise, which might well involve a different sort of interest in the spoon.

For the baby to have the opportunity to carry an interest and an action through to completion, a caring person or persons in a social context is necessary. Winnicott was observing not only babies, but babies with mothers. Babies need people who know that their development is important and who will try to understand their actions and the meaning of their actions. A caring social context remains crucial to our growth and development beyond infancy as well. One of the effects of trauma, however, is that the survivor's world of identification and caring shrinks. For the soldier in Vietnam this might have meant the sense that one's personal world shrank from an identification with one's country, or one's company, to a closeness with several buddies or only one, who might be killed as well. For other survivors this constriction of one's horizons might mean a shrinkage to immediate family, or a solitude perhaps combined with the search for just one trustworthy person to attach oneself to. The deathfulness within is manifested in the shrinking of the circle of care.

When death is loosed from the context of overarching meaningfulness and becomes intolerably final, the temptation—and the very widespread result—is to resort to a denial of death. This denial of death takes many forms, from a kind of manic activity accompanied by feelings of omnipotence to the deathlike apathy of "going through the motions of being alive." We see the denial of death in the soldier gone berserk, and we can see the berserk state as a form of death denial. As Jonathan Shay suggests in his book *Achilles in Vietnam,* "The berserker's manic obsession with revenge is not only destruction to gratify rage. At some deep cultural and psychological level, spilling enemy blood is an effort to bring the dead back to life."[20] Indeed, it is not so much death itself as the denial of death—in effect, not death so much as an end to birth—that is the antithesis of the concern for life.

Technology is the vehicle of our contemporary denial of death. While technology is essentially nonhuman and nonemotional, we identify with it, because the nonhuman is invulnerable to death. Identifying with the nonhuman, Searles would say, allows the terrified infant to feel omnipotent and not subject to the death that seems so terrifying. The terrified baby whose relation with the mother fails to provide an adequate sense of sheltering care, says Bion, rejects the possibility of emotional connection and opts for material comfort. But to opt for material comfort is to opt for the nonhuman, whose representative in the larger world is technology.

In a strange way, the soldier gone berserk displays a form of identification with the omnipotent machine. "The same human being may burn with fury while cold as ice, incapable of hunger, pleasure, or even pain from his wounds," notes Shay, who also notes how attractive the berserk warrior often is to those around him.[21] In the berserk warrior, the body itself has become the machine, cold as ice, but the emotion inside is fury, a rage to obliterate. To obliterate what?—the death and dying he has been immersed in. The psychoanalyst would explain the fantasy of omnipotence as something the baby creates to avoid the terror of helpless vulnerability. Is that not what the berserk soldier does, amid the overwhelming horror of death? In that warrior we see at one and the same time both the unbearable emotional pain and the fantasied solution of omnipotent destructiveness.

Contrary to the fantasied safety it promises, the denial of death amounts to a rejection of life, and it gives rise in turn to the fantasy of suicide, both as a way of actualizing the feeling of deadness and as an effort to escape by penetrating the barrier between life and death.

How Shall We Die? Suicide through Self-Neglect

> *Because we tend to feel that sudden death from nuclear warfare is a threat entirely out of our control, we may prefer the slower, more controllable death that pollution offers as seemingly the only alternative. . . . On the other hand, to regard such slow strangulation as an inevitable agony is to yearn for the quick relief that nuclear warfare would bring.*
> —Harold Searles, "Unconscious Processes in the Environmental Crisis"

If trauma and catastrophe render us unwilling to truly suffer and bear emotional pain, the pain we wish to evade may be not the terror of death itself so much as grief about overwhelming loss of life, whatever form that loss may take. From this point of view, the primary emotional work we need to do to deal with our inadequacy in the face of environmental destruction is the work of mourning. This means mourning for the destruction that has already occurred, in the hope that this work of mourning will free us to face the possibility of still greater destruction to come—destruction that may still be avoided if we put ourselves in the service of life.

We do not always think of trauma in terms of loss, but whether or not the actual deaths of individuals are involved, the trauma victim suffers the loss of social trust and the loss of the background security provided by the hope of symbolic immortality. These are catastrophic losses. And in fact, the trauma survivor shows the symptoms of what psychologists call morbid grief. Normal bereavement goes through a pattern of healing because it occurs within the context of an encompassing world fabric. But when loss overwhelms that fabric it results in a different picture.

Erich Lindemann, who made a psychological study of survivors of the catastrophic Coconut Grove fire in the 1940s, details the symptoms of morbid grief. They correspond closely to the effects of trauma. They include progressive social isolation, psychosomatic symptoms, restlessness and an inability to make decisions or initiate actions, and a loss of all feeling except for hostility, a kind of "going through the motions." Lindemann quotes from what he calls a typical report, from a young mother whose husband had died in the fire: "I go through all the motions of living. I look after my children. I do my errands. I go to social functions, but it is like being in a play; it doesn't really concern me. I can't have any warm feelings. If I were to have any feelings at all I would be angry with everybody."[22] Morbid grief may also take the form of what is called "agitated depression," acute feelings of worthlessness and self-accusation that may lead to suicide. We have already heard about what Lindemann calls "self-dispossession," in which the person overwhelmed by the encounter with death gives away all he or she has, to the point of self-destruction. Searles suggests that much of the irrational and destructive behavior we see could be understood as a suicidal depression of this kind.[23] In suicidal depression, we look to dying as a way of simultaneously dealing and not dealing with overwhelming loss, in much the same way as the soldier gone berserk becomes simultaneously death dealing and death defying.

Nuclearism and environmental degradation, Searles suggests, seem to both raise and answer the question, How can we die in such a way that we won't feel the pain of it? It is as though the losses we have already suffered impel us to more extreme destruction, to the point of Lindemann's self-dispossession or Searles's suicide through self-neglect. Both the berserk and the suicidal answers might then run something like this: We could identify with an all-powerful technology so that we don't have to experience the perils of our limitations and mortality, and we could get it all over with through a nuclear war so that we don't have to experience a slow suffocation through pollution. Alternatively, we could ruin our earth so that we can see before our eyes the destruction we have experienced internally, and die the death of those who have lost something valuable, rather than ending our days as nonhuman robots in a colony in outer space.

Survival, Guilt, and Complicity

> *No survival experience . . . can occur without severe guilt.*
> —Robert Jay Lifton, *Death in Life*

> *The authorities that instituted the ritual sacrifice of the Bouphonia of ancient Athens devised it so that no one could be held responsible for the violent decapitation of an ox. . . . They chose girls to bring the water used for sharpening the axe and knife. After sharpening, one man administered the axe, another struck the ox, and a third cut its throat. The next skinned it, and everyone tasted the ox. Afterwards having sewn up the hide of the ox, they stuffed its skin with hay, sewed it up and had it stand yoked to a plough, as though it were ready for work. . . . They summoned all who had participated in the deed to defend themselves. The water fetchers charged that the sharpeners were more to blame than they. The sharpeners said the same about the axe-administrator, and this one of the throat cutter and this one of the knife which, being without a voice, was condemned to murder. . . . They threw the knife into the sea. (Porphyry,* de abstentia)
> —Eric Rhode, *On Hallucination*

In the current situation of ongoing environmental destruction, those of us who live in the West (and certainly some who live elsewhere) are all in some degree both perpetrators and victims. We are likely to be enjoying what others—both in the West and elsewhere—would perceive as affluence. We may also be among those who make decisions that directly affect the survival of species and cultures. All of us, I think, sense this at some level. We know we are caught up in a system that causes damage and must cause further damage simply in order to continue operating. At the same time, we know that we and others are suffering from that damage, to whatever degree, and that we are all likely to suffer still more directly and more painfully in the future. On both counts, we feel guilty. How can we do more? How could we have participated in doing what we have already done? How can we live knowing that we are part of this impossible situation? "Grief . . . can lead men to give up all desire to return home alive. The searing guilt of 'it should have been me' has the same result," writes Shay.[24]

In his essay "Surviving," Bruno Bettelheim makes the important point that in the concentration camps individuals could not ensure their own survival. What prisoners might or might not do might or might not help them stay alive. For most people nothing ensured survival; they perished. Release came, if at all, only from above or outside, virtually never from the individual's own efforts. The notion that we do well to exchange moral integrity for "survival" is thus a delusion, Bettelheim argues. To the contrary, having something to live for at least kept some people alive long enough to be freed by the ending of the war.[25] Indeed, complicity contributes to the evil of the situation. First, the lack of resistance encourages the escalation of atrocity. When the Nazis early on attempted to exterminate the mentally ill, for instance, the immediate protest by family members quickly led to that attempt being dropped.[26] Second, the lack of resistance in others intensifies the feelings of helplessness and abandonment in the victims, lowering, as it were, the life orientation of those involved in the death-producing situation.

If it makes no sense to abandon one's moral integrity for survival, still, in what Lifton calls "atrocity-producing situations," which involve the large-scale social administration of violent and premature death in some form, all of those involved, whether primarily victims or primarily perpetrators, absorb the evil of the situation as a whole. The survivor of such a situation—and as Lifton points out, all who are alive now are survivors, not only of Hiroshima but of the ongoing possibility of worldwide destruction—"turns his anger inward precisely because he cannot help but accept and internalize the world in which he has been victimized, including in some degree the motivation and behavior of the victimizer."[27] We feel guilty, knowing that we are tainted in this way.

Guilt is an experience of failed responsibility, and failure is inevitable in these situations. "In a disaster of this magnitude," Lifton explains, "the extreme conditions drastically limit the possibilities of cooperation and mutual aid, and thereby greatly accentuate the awareness of ordinary urges toward self-preservation. The idea that an individual's first and strongest impulse is directed toward his own survival becomes vividly displayed and, in this death-saturated context, totally unacceptable. Even more unacceptable is the inner joy at having survived, whatever the fate of one's fellows."[28] Thus guilt affects everyone who is responsible for helping others in a situation that makes giving adequate help impossible. After the bomb had been dropped on Hiroshima, for instance, those who were still alive were surrounded by innumerable others crying for help, often including members of their own family. Hiroshima survivor Takashi Nagai articulates this painful sense of guilt: "Those who survived the bomb were, if not merely lucky, in a greater or lesser degree selfish, self-centered, guided by instinct and not civilization. . . . and we know it, we who have survived. Knowing it is a dull ache without surcease."[29]

Some of the most painful memories that survivors carry are of their failures to help in these circumstances. Randy Morris tells the story of going to Hiroshima as a naive young man to collect the dreams of *hibakusha* (A-bomb survivors). One survivor told him this recurrent nightmare:

> You see in this dream I was twelve years old again and I woke up on a fine morning. I was walking to school when this big bomb exploded all around me. I was screaming and my dress was on fire, as was the land all around me. People were yelling and running every which way. I tried to make it to this bridge that would lead me back home. As I crawled over the bridge I heard my name being called: Tomoko! Tomoko! I looked into the water where all these bodies were floating, and there I saw my friend Akiko. She was still alive, slowly turning in the current. Save me, Tomoko, save me! Ah, but Randy-san, you see I was too weak to do anything. I couldn't save her. I turned away and walked on. And then I wake up feeling terrible!

And then the survivor turns to Randy and tells him that this is what really happened. She dreams, over and over again, what really happened. She really was too weak to help, and she never saw her friend again.[30] The experience cannot become a real dream; she cannot recover from the guilt of impossible weakness. We see a version of this guilt in people whose environmental concern is awakened by their concern for their children's future. What awakening concern may put them in touch with, ironically, is not a sense of efficacy so much as this sense of urgent responsibility for something that far outstrips their objective capacity to help.

Guilt is painful and hard to bear. The pressure to put an end to it is enormous. That pressure easily leads the person to increasing complicity with destruction, and to denial of guilt later on. In his essay, "The Psychological Appeal of Totalitarianism," Bettelheim recounts the sad story of people who opposed the Nazi regime but for one reason or another had something they did not want to give up. They found themselves in a conflict, an untenable situation. They had to conform outwardly to some extent in order to survive and hold onto the things they did not want to give up. But since Nazism demanded conformity not simply on the outside but on the inside as well, it was difficult to conform outwardly while maintaining the inward resistance. There was always a question of where to draw the line in order to be safe. In virtually all these cases, Bettelheim reports, the person eventually resolved the unbearable cognitive dissonance by becoming a supporter of the regime.[31]

Because guilt cannot be resolved without suffering the intense pain it involves, attempts to deny it internally and externally can last a lifetime. When Lifton decided to write about the psychological situation of the medical doctors who participated in the Nazi killing programs, he expected to find the surviving doctors reluctant to speak with him. He carefully secured letters of introduction that he thought might make the doctors more amenable to being interviewed. To his surprise, he found most of the doctors he approached quite willing to be interviewed. To his even greater surprise, he found that they felt neither guilty nor remorseful. Instead, they were interested in being interviewed in order to present themselves as having become good people once the war was over. They had denied their guilt by dissociating themselves from the persons they were during the Nazi period.[32]

Similarly, as part of his work on nuclearism, Lifton has investigated the decision to use the atomic bomb in Japan when alternatives were available, and has traced the way the official narrative has evolved to legitimize that decision. *Hiroshima in America,* the book he wrote with Greg Mitchell as a contribution to the nation's reflections on the fiftieth anniversary of Hiroshima in 1995, details the case of Harry Truman, the president who had final responsibility for the decision. It follows Truman as he becomes adamant in his denial of guilt and in his dismissal of alternatives, and traces what Lifton and Mitchell call his "struggle against authentic witness"—his unwillingness to give in to his own impulses to confront and reflect on his actions, the pressures brought to bear on him, and the consequences of his decision.[33]

If denying one's guilt contributes to the deathfulness of the situation, it also keeps the person in a state of psychological stasis, a form of the "death in life" that is the subject of Lifton's book on the survivors of the A-bomb in Hiroshima. In contrast to this kind of static guilt, as Lifton calls it, is what he terms *animating guilt.* In animating guilt, one brings oneself back to life by feeling and reflecting on one's guilt, as Truman was never able to do. Lifton quotes a Vietnam deserter's life-restoring insight about his guilt: "I was somebody with feelings who had done something wrong and I—I was not an animal or some kind of killing machine. . . . I physically felt satisfaction, knowing that I felt guilt. Meaning that I was not a murderer."[34] Animating guilt involves not punishing oneself so much as wrestling with the existential questions of what guilt is, what it means to be human, how personal responsibility and the nature of the atrocity-producing situation were both involved in the person's experience. The deserter realized that experiencing and suffering his guilt allowed him to be human and alive, whereas denying it would mean being nonhuman, a mere killing machine—his definition of a murderer.

In many cases, the survivor's personality has partially or completely disintegrated under the impact of the traumatic experience. The survivor has in fact abandoned his or her moral integrity under the unbearable pressure of what Bettelheim termed the *extreme situation.* Bettelheim outlines the survivor's options: He can despair and give up hope of finding reintegration; he can try to repress and deny the experience in the hope of getting back to "things the way they used to be" and becoming again the self he used to be—an understandable but useless reaction; or he can seek reintegration of a new sort, taking into account the burden of what he has experienced and the responsibility for making meaning of it.[35]

What would making meaning of such an experience consist of? It would include facing and coming to terms with the fact that one has survived where others have not, as well as making meaning of the double guilt of one's failed responsibility and one's joy at having survived. The reintegration that comes from such an effort would include the experience of animating guilt described above. It might include as well what Lifton calls a *survivor mission,* where one deals with one's experience by devoting oneself to helping others understand and to preventing a repetition in the future.[36] Hence some of the atomic scientists became antinuclear advocates, and some of the *hibakusha* in Japan became spokespeople for peace. In such a mission the immediate personal experience of large-scale killing and dying is of value. The survivor can bring this awareness to the larger community. If, as Lifton points out, we are all now survivors of Hiroshima, as well as survivors of environmental destruction, perhaps some of our efforts represent our own survivor mission.

Already Dead: Apathy and the Mask of Suffering

> But I do not really know whether I have survived? My inner self has shut itself up more and more. As though to protect itself, it has become inaccessible even to me, and so I do not know now whether in my heart's core there is still the strength to venture upon world-relationships and realize them, or whether only a tombstone of my former spirit has quietly remained there.
> —Rilke, Wartime Letters of Rainer Maria Rilke (1919)

"Most people can endure only a certain number of disasters at a certain rate of speed before they pass into a state of demoralization, in which they are practically incapable of initiating anything, although they are able to keep walking, to maintain routines, and to carry on customary tasks," wrote Harry Stack Sullivan in The Psychiatric Interview.[37] That was in the 1940s, and at that time, Sullivan noted, one did not ordinarily see many people in that condition in the United States. It was in some branches of the military, or in some areas of war-torn countries that one had "an excellent opportunity to become acquainted with apathy of all degrees and grades."[38] Since then we have all had the opportunity to become intimately acquainted with apathy.

Apathy, Joanna Macy reminds us, is the mask of suffering.[39] A hard shell, that is, with which to meet, or appear to meet, a world too disturbing to make real contact with. Apathetic people do not show interest, get engaged, get involved. The mask stands between them and whatever it is they might get involved in. To show interest means to be in the middle of something, not behind a mask. Until the apathetic person has enough sense of a larger sheltering to take the place of the masklike protection, a sheltering from which to venture forth, confident of a safety to which to return, he or she will stand apart.

"Apathy seems to me," writes Sullivan, "to be a miracle of protection by which personality in utter fiasco rests until it can do something else."[40] It is a way of adapting, of defending oneself in a situation that is utterly overwhelming and where there is no end in sight. We all need this kind of protection in our current environmental situation, certainly, in which there is so much destruction of so many kinds, in so many places, affecting so many people and so many other creatures, and with no end in sight. No wonder environmental activists complain about the widespread apathy that meets their efforts to arouse concern.

In the soldier in combat, apathy represents a sense of having already been killed. The encounter with overwhelming deathfulness that destroys the background fabric of the world and obliterates the hope of some kind of symbolic immortality, plunges one into a frame of mind in which one is already dead. "The sense of being already dead may contribute to the berserker's complete loss of fear," writes Shay in *Achilles in Vietnam*. But as we have seen, the berserk state is a denial both of life and of death—a death in life, as Lifton titled his book on the survivors of Hiroshima, a denial of mortality by denying participation in the process of living. The sense of already being dead, Shay continues, "may also be the prototype of the loss of all emotion that defines for combat post-traumatic stress disorder the prolonged states of numbness—the inability to feel love or happiness or to believe that anything matters."[41] But apathy is not a willful act. It is the result of an inability to believe that anything matters—that anything can counterbalance the overwhelming deathfulness in the scales of reality. Trauma obliterates hope and desire.

Apathy signals an inability to believe that the boundary between life and the situation of death-immersion can be crossed again. "In my wildest thoughts I never expected or wanted to return home alive," Shay quotes one Vietnam veteran as saying, "and emotionally I never have."[42] The narrator in Erich Maria Remarque's novel of the First World War, *All Quiet on the Western Front,* describes his memories—or are they daydreams?—of the city where he used to live, scenes of the poplars lining the river, of the cathedral. How still those scenes are, he muses, compared to the endless noise of the front—and how unutterably and hopelessly divorced and distant they are from him in their stillness. He could never hope to reconnect with that distant life.[43]

This sense of being caught in a war without end, and the sense that our experience of that war separates us utterly from those without that experience, is both the unconscious suffering of those who have become apathetic, and the frustration of those who feel an urgent concern and try to rouse the others. How many of us can really imagine that the war against nature will be over and we will come out alive in a world where continuing ecological destruction is not the order of the day? (I remember the moment when it suddenly and surprisingly occurred to me that the war in Vietnam could not, would not last forever, a moment of utter surprise, joy and relief.) It behooves us to remember that apathy is indeed the sign of a personality—or a society—resting in the midst of utter fiasco, and to ask ourselves when and under what conditions it might be able to reemerge.

Numbing and Doubling

> In talking with a naval officer who had won a target competition with big naval guns, we asked, "Have you thought what happens in a little street when one of your shells explodes, of the families torn to pieces, a thousand generations influenced when you signaled Fire?" "Of course not," he said. "Those shells travel so far that you couldn't possibly see where they land." And he was quite correct.
> —John Steinbeck and Edward F. Ricketts, *Sea of Cortez* (1941)

Robert Jay Lifton developed the term *psychic numbing* to describe the way we shut off awareness of the destructive consequences of our acts, thus preventing ourselves from responding emotionally to the pain we cause. We do not hear the catfish scream, we do not imagine the consequences of the bombs we drop. Numbing is a form of dissociation. We split off one part of ourselves from another. It is a form of psychic distancing, a denial of intimacy, as when we dissociate ourselves from a former comrade. We see numbing in extreme form in direct acts of obvious destructiveness. It is less extreme, less dramatic, and less noticeable in the rest of us. We need to numb ourselves simply to get through the day. "Within this culture," Lifton says, "you could call the numbing of everyday life a necessary defensive maneuver."[44] Our technology-saturated world both exacerbates our need for numbing and at the same time offers us numbing in handy form. It does so by bringing what is literally distant closer to us in a form that is prenumbed, as it were, as in the wildlife documentary in which the taste, smell, temperature and space of the habitat have been excised. And it does so by allowing us to take actions that have their destructive effects at great spatial (and often temporal) distances from us, as with the naval officer Steinbeck and Ricketts spoke to in the early 1940s. If the officer were not numb, Steinbeck and Ricketts write, "if he could really see where [the bombs] land and what they do, if he could really feel the power in his dropped hand and the waves radiating out from his gun, he would not be able to perform his function. He himself would be the weak point of his gun."[45] Full physical and emotional participation, that is, are at odds with the technological function. "[Technology] intensifies every form of dissociation," Lifton writes. Even an ostensibly benign technology enhances dissociation, he notes, "by distancing decisions not only from the consequences themselves but also from awareness, psychological perception, and psychological response to the consequences."[46] Indeed, we become so used to such dissociation that we lose sight of the complex wholeness that characterizes natural actions.

A still more extreme form of dissociation is what Lifton calls *doubling:* "the formation of a second self-structure . . . morally at odds with the prior self-structure."[47] Doubling allows the person, via a second self or double, to participate in actions morally reprehensible to the first self. In other words, someone can be one person, morally speaking, in a destructive activity and another person, morally speaking, in another area of life, as perhaps with family, or as athletic coach or patron of the arts. We are all familiar with this from stories about individuals who have committed horrifying crimes: "Mr. X, father of five, churchgoer, teacher . . . and serial murderer."

Doubling becomes widespread in situations where large groups of people are both victims and perpetrators. If and when the destructive operations come to an end, or the person somehow dissociates himself or herself from them, the person may return to the original self and feel alienated from and amazed by the existence of the second self. Doubling has been studied in clearly demarcated professional groups that participated in large-scale destruction: the medical profession in Nazi Germany with its almost wholesale complicity with the killing program, or the atomic scientists in the United States during and after the Second World War. "I would like to think now, that at the time of the German defeat, I would have stopped and taken stock, and thought it all over very carefully, and that I would have walked away from Los Alamos at that time," says Robert Wilson, a scientist involved in the creation of the atomic bomb. He is mystified by his failure to act: "In terms of everything I believed in before and after and during the war, I cannot understand why I did not take that act."[48] Professional socialization and bonding with colleagues in conditions of secrecy, as was the case at Los Alamos, play a crucial role in the formation and maintenance of the second self. Wilson notes that he cannot recall a single instance of one of his colleagues walking away, or even suggesting that possibility. "Our life was directed to do one thing," he continues. "It was as though we had been programmed to do that and as automatons were doing it."[49]

In describing himself and his colleagues as automatons, Wilson speaks from a self that no longer shares the motivations of the Los Alamos scientist. But as Lifton points out, doubling always involves a claim to virtue, and we cannot understand it if we ignore that.[50] There is a claim that the destructive operations are in the service of some shared value—"purification," "renewal," "saving American lives," "providing clean energy," "creating jobs," and so on. This is as true in the subtle numbing involved in environmental destruction, Lifton claims, as in the Nazi ideology. The professional collegiality that helps to create and maintain the doubled selves rests on these shared values. It is simply that a more spacious self would balance and integrate these values with others.

The second self, then, tends to live in a narrow moral world. A move out of the viewpoint of the second self into greater spaciousness of thought would be experienced as threatening the whole psychic world within which the second self lives. "The military mind must limit its thinking to be able to perform its function at all," write Steinbeck and Ricketts, reflecting on the gunnery officer they talked with. " [He] is too humble to take the responsibility for thinking. The whole structure of his world would be endangered if he permitted himself to think," they conclude. "By not seeing, by insisting that it be a problem of ballistics and trajectory, he is a good gunnery officer."[51]

The phenomenon of widespread doubling raises difficult questions. Is it possible to be morally consistent in the situations in which we live? The attempt to be morally consistent may involve an identification with the apocalyptic self, which is in turn tempted by dogma and the fantasy of a "regeneration through violence." As an alternative, Lifton has put forward some concepts that get at the kind of spaciousness I invoked above: a "protean self," in which one patches together pieces of goodness and takes a more flexible approach to different situations, oriented by some core principles larger than the individual self, and a "species consciousness," a sense of humanity that would transcend individual group identifications.[52] The notion of an "ecological consciousness" discussed earlier moves in the same direction. Can socialization be directed toward identifications like "species consciousness"? Can group process be life enhancing?

The Collective and the Unquestioned: How Ordinary People Participate in Horrifically Destructive Acts

> *Is the growth of our wisdom likely to keep pace with our intelligence?—using intelligence in the pejorative sense of the ability to learn tricks. It is a matter of the greatest possible urgency that the human animal should discover what sort of animal he is before he has blown himself off the earth. The monkey-trick department is far and away in advance of the rest.*
> —Wilfred Bion, *Clinical Seminars and Other Works*

Lifton's notion of doubling and dissociation reminds us that individually we are generally moral people. Within our limitations, we attempt to do good and to act decently. Nevertheless, whether we feel it directly or only betray it in the urgency with which we reject attempts to bring it to our attention, we live with a painful sense of guilt for participating, however involuntarily, in the ongoing destruction. How is it that we, who are not bad people, come to share in the guilt of such horrible things?

In an interview for his collection *Listening to the Land,* Derrick Jensen asked Lifton this very question: "How is it that ordinary people participate in horrifically destructive acts?" Lifton's reply is instructive. "Most evil—or destructiveness, which is almost the same thing," he says, "emerges from shared ideology and practice that tends to be unquestioned." Most of the people who engage in potentially destructive projects, he continues, are in fact ordinary, and these ordinary people "engage in profoundly immoral acts through ideology, group process, unquestioned personal decisions, and the group, national, or corporate behavior and policies they become part of."[53] Participation in the group's project and its ideological justification can obscure one's own ethical thinking, and lead one to ignore the consequences of one's actions.

Lifton's remarks ring true. It is not enough to reflect on ourselves as individuals. If we are to understand not only destructiveness but also our resistance to becoming aware of organized—that is, ideologically justified—destructiveness, we have to understand the psychology of groups as well. It is Wilfred Bion who, to my mind, has done the most illuminating thinking about groups.

For Bion, as for Lifton, there is no escaping the group. Human beings are political animals who cannot exist without groups, though they are also individuals. The problem is that though groups are often capable of productive work on the task at hand, they are also permeated by a strange irrationality that Bion terms the *group mentality*. Think of the conflicts we all feel about groups. On the one hand, we want to be "part" of the group, which generally means in some way being part of something larger than ourselves that we do not question. On the other hand, we want to retain our sense of individuality, our capacity to look from the outside and exercise our critical capacities and our judgment. We can never be satisfied. When we are part of the group we feel swallowed up. When we assert our individuality and stand on the outside, we feel precisely that—an outsider, lonely and peripheral. As Bion puts it, "man is at war with his groupishness."[54]

The group mentality Bion speaks of refers to the irrational state of mind in which we are part of, indeed one with, the group. Groups behave as though under the sway of what Bion calls *basic assumptions*—assumptions, essentially fantasies, that are contrary to reality but nevertheless seem to be the basis for the group's actions. Bion identified three of these fantasy-based states of mind. One is *dependency,* a condition in which all members of the group seem to be passively waiting to be taken care of by the ostensible "leader," who is regarded as an all-wise, if terribly withholding, god. Another fantasy-based state of mind is *fight-flight,* in which it seems as though there are only two alternatives to some terrible threat, namely, to flee all together or to fight to the death, either of those to be undertaken at the drop of a hat or the merest whisper of the commander. Finally, there is what Bion calls *pairing,* in which the group watches with fascination the interaction, imagined to be sexual, between two members, an interaction that in the minds of all will produce something new and great—a new idea, a utopia, a messiah of some sort.

Characteristic of all these fantasy-based group mentalities is a lack of open mental space, so to speak, for differentiation; the group mentality is characterized by uniformity of sentiment, immediacy of action, and lack of thought or disagreement. It does not stop to consider consequences. To the extent to which the group's actions are dictated by the group mentality, they are not oriented to reality but are rather expressions of the shared mentality. In these terms, what Lifton calls ideology would be a rationalization of one or another of these governing fantasies, and what Bion calls groupishness would militate against questioning the consequences of acting on that ideology.

In the early 1970s Irving Janis published a book called *Groupthink,* which detailed instances of something very like this—instances in which terrible political decisions were reached by groups of very intelligent and well-informed men. In a group operating under the sway of "groupthink," Janis points out, criticism is squelched—not only out of a fearful "respect" for the leader but also through active policing on the part of some members to make sure that alternative views are discredited even before being considered.[55] It is easy to see how groupthink would have an appeal in the face of the dilemmas posed by environmental degradation. What a relief to be part of a group with a "strong leader" who will presumably take care of everything. What a relief to have a clear-cut enemy to fight against, or something horrible to rid one's life of. What a relief to be one of a group of worshippers of a brilliant and creative pair expected to come up at any moment with a wonderful idea that will provide (usually technological) solutions to pressing problems.

Though groupthink is based in fantasy, the destruction it leads to is real, however little the participants notice it at the time, so caught up are they in the activities peculiar to the fantasy dominating the group. The damage includes not only the neglect of real issues and the difficult search for genuine alternatives but damage to "enemies" within and without. It entails the erosion of the capacities of those in the dependent group, and the tacit squashing of creativity in the pairing group in order to make sure that messianic hope can continue to fuel the group in its search. The effects of groupthink can be seen in the frequency of scapegoating and "demonizing," in the rapid turnover of leaders, in rivalries between groups working toward similar ends, in the preoccupation with the private lives and loves of the prominent, and in the ways all these impede the tasks that members have, often correctly, set for the groups. Activist groups are in no way immune to these dangers.

The Environment as Scapegoat and Sacrifice

> *There's a huge rectangle which embraces parts of four States—Utah, Colorado, New Mexico and Arizona—and which is nothing but enchantment, sorcery, illusionismus, phantasmagoria. Perhaps the secret of the American continent is contained in this wild, forbidding and partially unexplored territory.*
> —Henry Miller, *The Air-Conditioned Nightmare* (1945)

Following René Girard, Gil Bailie claims that every culture and civilization to date has been founded on a violent sacrifice. That foundational sacrifice is then

enshrined and beautified in myth and religion. The tacit assumption has been
that the group cohesion achieved through scapegoating confers culture-building
benefits—order, stability, and the like. At some point, however, the efficacy of
the foundational sacrifice wears thin, and a new sacrifice is required. At the current stage of civilization, Bailie argues, we have gained enough awareness of
the perspective of the victim (through Christianity, he would say) that the efficacy of new violent sacrifices is wearing thin. Repentance comes almost immediately, and the participant wonders what on earth he was doing. Hence, Bailie
argues, we are now in a unique position: We need to unveil further this violence at the heart of our culture and learn to live without using violent sacrifice. If not, we will descend into more and more violence that is less and less
efficacious in creating order. We are already seeing this in the massacres and
civil wars occurring around the world.[56]

The Four Corners area of the Southwest has been called a "national sacrifice
area." Perhaps the "secret of the American continent" that Henry Miller sensed
in that area[57] is the secret of our founding sacrifice of the mystery and wildness
of the natural world. Perhaps we imagine that through that sacrifice we are putting the problem of our continuity and discontinuity with the rest of the nonhuman world out of our minds and making a sacrifice that will bond the
human group together. If there is no wild left, we will not need to ponder the
contrast between what is wild and what is not. Certainly the West has long
been bonded as a "civilization" through the notion of the domination of nature.
Perhaps the dynamic of increasingly less efficacious violence that Bailie predicts
is now at work in our increasing destructiveness toward the larger natural environment, the increasingly flamboyant and omnipotent destructiveness of multi-
national corporatism. But perhaps now too, as Bailie suggests, we are more
able to take the perspective of the victim, and thus we have intimations that at
some point we will "come to our senses" and ask what on earth we thought we
were doing.

A Place in the Mind: Survivor Illumination and the Difficult Knowledge of Death

> *Western philosophy has long realized that we must imagine our own death
> in order to live more fully. After Hiroshima, however, our further task is
> to imagine the end of the world in order to take steps to maintain human
> existence.*
> —Robert Jay Lifton and Greg Mitchell, *Hiroshima in America*

The current situation calls for what Wilfred Bion calls *binocular vision:* we must pay attention both to the possibility of catastrophe and to the alternatives. If we do not face the genuine possibility of further catastrophe, we cannot envision a response. This is why Lifton and Mitchell caution us to imagine the end of the world. One reason we do not let ourselves imagine such a possibility, however, is that we have, as Lifton points out, "no preexisting place in the mind" for it.[58] Even those who have survived large-scale catastrophes cannot always accept that they really happened. Yet survivors have the possibility of playing a tremendously valuable role here. For as Lifton points out, though they may close down as a result of their experiences, they may also open out. This he calls *survivor illumination:* they can bring an awareness to us based on their "knowledge of death"—their contact with the possibility of killing or dying on a massive scale.

As I pointed out earlier, everyone living now is a survivor not only of the nuclearism that is a legacy of Hiroshima and Nazism but also of massive environmental destruction around the world. Those catastrophes have occurred and continue to occur in the world that is our world. Yet our own experience as survivors reminds us that the closer one comes to awareness of apocalyptic possibility, the more difficult it is to retain a mind open to multiple possibilities. Going berserk and yielding to unrestrained destructiveness becomes a temptation. "Restraint is always in part the cognitive attention to multiple possibilities in a situation," notes Shay. "When all restraint is lost, the cognitive universe is simplified to a single focus. The berserker is figuratively—sometimes literally—blind to everything but his destructive aim."[59]

The temptation to close one's mind and destroy new thoughts and questions that might want to enter is one form of berserkness. The closed mind seeks dogma. "When one contemplates the threats we face, which involve apocalyptic dimensions of killing or destroying," Lifton comments, "one is likely to call forth dogma."[60] Recourse to dogma is an understandable attempt to put an end to the anxiety of uncertainty about the future, and to do so through submission to authority. It can represent a fantasied protection against the pain of guilt as well. If someone would only tell us what to do, we imagine, we could evade the risk of feeling guilty for the consequences of actions taken in uncertainty. "Human beings clamour for some kind of authoritative statement to take the place of both ignorance and the exercise of curiosity," Bion reminds us. They "hope, in that way, to put a stop to disagreeable feelings of ignorance and the repetition of the questions."[61]

The place in the mind that Lifton talks about can be thought of as the spaciousness of a mind that can contain both pain and curiosity. Is it possible, we might ask, to be both curious and disturbed about the environmental situation? How strange this question seems. But why does it seem so strange to think of being curious about the environmental situation? I think it is because the situation is so alarming that we cannot imagine turning our attention to it without losing our capacity for detachment and inquiry. Could anyone who was not abnormally detached simply be curious about environmental destruction, we wonder. On the other hand, we are well aware how much we wish we were not so disturbed about the situation. (Why do I care so much, a friend of mine asked, partly in curiosity, but partly in lamentation.)

Still, if we think of the dangers of denial and apathy (admitting that they can have a certain partial or temporary usefulness), it is clear that there is value to being able to be disturbed. Despite the strain it imposes on the psyche, it is important to be able to respond to something disturbing by being disturbed. Similarly, it is clear that the capacity to be curious is very useful. Since the environmental situation is a part of current reality, albeit a disturbing part, and since it is something we do not yet understand or know how to deal with, it is useful to be able to turn our attention to it, to notice how disturbing it is, and yet to be interested in it and inquire into it.

Let me propose, in fact, that the capacity to be curious *depends on* a simultaneous capacity to be disturbed. For curiosity cannot receive a genuine response to what it investigates if the person cannot tolerate emotional disturbance. As Bion remarks, "The intolerance of frustration, the dislike of being ignorant, the dislike of having a space which is not filled, can stimulate a precocious and premature desire to fill the space."[62] What precocious and premature desire might fill the space with will not be something thoroughly felt and understood. It cannot lay the psychological groundwork for alternatives to the end of the world.

Notes

1. Michael Ortiz Hill, *Dreaming the End of the World,* 89.

2. See Robert Jay Lifton and Greg Mitchell, *Hiroshima in America,* 343.

3. Rainer Maria Rilke, *Wartime Letters of Rainer Maria Rilke,* 110.

4. See Lifton and Mitchell, *Hiroshima in America,* 352–353.

5. Hill, *Dreaming the End of the World,* 68.

6. Hill, *Dreaming the End of the World,* 3–4.

7. Arthur Colman, *Up from Scapegoating,* 90.

8. Lifton, *Death in Life,* 541.

9. For an excellent general presentation of the psychological impact of trauma, see Judith Lewis Herman, *Trauma and Recovery.*

10. Bruno Bettelheim, "Schizophrenia as a Reaction to Extreme Situations," in *Surviving, and Other Essays,* 115.

11. Harold Searles, "Unconscious Processes in the Environmental Crisis," in *Countertransference and Related Subjects,* 237.

12. See Searles, "Unconscious Processes in the Environmental Crisis," in *Countertransference and Related Subjects,* 232.

13. Jacques Lusseyran, *Against the Pollution of the I,* 152.

14. Searles, "Unconscious Processes in the Environmental Crisis," in *Countertransference and Related Subjects,* 233.

15. Lifton, *Death in Life,* 486.

16. See Searles, *The Nonhuman Environment in Normal Development and in Schizophrenia,* esp. 9–11, 39–41.

17. Searles, "Unconscious Processes in the Environmental Crisis," in *Countertransference and Related Subjects,* 239.

18. Lifton, *Death in Life,* 484.

19. See D. W. Winnicott, "The Observation of Infants in a Set Situation," in *Through Paediatrics to Psycho-Analysis,* 52–69.

20. Shay, *Achilles in Vietnam,* 89.

21. Shay, *Achilles in Vietnam,* 93.

22. Erich Lindemann, *Beyond Grief,* 70.

23. Searles, "Unconscious Processes in the Environmental Crisis," in *Countertransference and Related Subjects,* 239.

24. Shay, *Achilles in Vietnam,* 73.

25. Bettelheim, "Surviving," in *Surviving, and Other Essays.*

26. On resistance to the Nazi "euthanasia" programs, see J. Noakes and G. Pridham, eds., *Documents on Nazism, 1919–1945,* and R. Proctor, *Racial Hygiene.*

27. Lifton, *Death in Life*, 497.

28. Lifton, *Death in Life*, 47.

29. Lifton, *Death in Life*, 48.

30. Walter Enloe and Randy Morris, *Encounters with Hiroshima*, 12–13.

31. See Bettelheim, "The Psychological Appeal of Totalitarianism," in *Surviving, and Other Essays*.

32. Lifton, *The Nazi Doctors*, 6–8.

33. See Lifton and Mitchell, *Hiroshima in America*, Part II, "Harry Truman's Tragedy."

34. Lifton, *Home from the War*, 128.

35. See Bettelheim, "Trauma and Reintegration," in *Surviving, and Other Essays*.

36. See Lifton, *Death in Life*, esp. 302–305.

37. Harry Stack Sullivan, *The Psychiatric Interview*, 188–189.

38. Sullivan, *The Psychiatric Interview*, 174–175.

39. Joanna Macy, *World as Lover, World as Self*, 191.

40. Sullivan, *The Psychiatric Interview*, 174–175.

41. Shay, *Achilles in Vietnam*, 53.

42. Shay, *Achilles in Vietnam*, 53.

43. Erich Maria Remarque, *All Quiet on the Western Front*, 105–107.

44. Lifton, interview in Derrick Jensen, *Listening to the Land*, 143.

45. John Steinbeck and Edward F. Ricketts, *Sea of Cortez*, 40–41.

46. Lifton, in Jensen, *Listening to the Land*, 143–144.

47. Lifton, in Jensen, *Listening to the Land*, 145.

48. In *The Day After Trinity*, Pyramid films, 1981, quoted in Michael Ortiz Hill, *Dreaming the End of the World*, 32–33.

49. In *The Day After Trinity*, Pyramid films, 1981, quoted in Michael Ortiz Hill, *Dreaming the End of the World*, 33.

50. Lifton, in Jensen, *Listening to the Land*, 144–145.

51. Steinbeck and Ricketts, *Sea of Cortez*, 40–41.

52. See Lifton, in Jensen, *Listening to the Land*, 151, and Lifton and Markusen, *The Genocidal Mentality*, chap. 9, "A Species Mentality."

53. Lifton, in Jensen, *Listening to the Land*, 142.

54. See Wilfred Bion, *Experiences in Groups,* for his ideas on groups. For Bion's discussion of the conflict between the individual's sense of separateness and his identification with the group, see for instance p. 131 in that volume.

55. Irving Janis, "Groupthink," in David Kolb, Irwin Rubin, and James McIntyre, eds., *Organizational Psychology,* 3rd ed., 236–246, esp. 243.

56. See Gil Bailie, *Violence Unveiled.*

57. Henry Miller, *The Air-Conditioned Nightmare,* 239.

58. Lifton, in Jensen, *Listening to the Land,* 150.

59. Shay, *Achilles in Vietnam,* 86.

60. Lifton, in Jensen, *Listening to the Land,* 147.

61. Bion, "Emotional Turbulence," in *Clinical Seminars and Other Works,* 302.

62. Bion, "Emotional Turbulence," in *Clinical Seminars and Other Works,* 301–302.

6   The Future and the Possible

The fundamental problem is, how soon can human beings reconcile themselves to the fact that the truth matters?
—Wilfred Bion, *Attention and Interpretation*

After the final no there comes a yes.
And on that yes the future world depends.
—Wallace Stevens

Thich Nhat Hanh titled an anthology of writings on the Buddhist precepts of mindfulness *For a Future to Be Possible*. This title takes note of the intense anxieties we have about the future. The fact that Thich Nhat Hanh's title surprises us, however, points up the way we assume that the future rolls along automatically, removed from our participation. His title implies that the future is in question. Can it be possible that there will not be a future? That idea strikes us as on the one hand absurd and on the other hand as unacceptably and terrifyingly plausible. What can it mean?

Thomas Berry speaks of this future with the penetrating vision of his old age: "Our present system, based on the plundering of the Earth's resources, is certainly coming to an end. It cannot continue. The industrial world on a global scale, as it functions presently, can be considered definitively bankrupt. . . . In the future the industrial system will have its moments of apparent recovery, but these will be minor and momentary. The larger movement is toward dissolution."[1]

"It is a matter of the greatest possible urgency that the human animal should discover what sort of animal he is before he has blown himself off the earth," wrote Wilfred Bion.[2] "A matter of the greatest possible urgency"—Bion's phrase seems correct. When we think of the direction of the larger world, we sense urgency. Something seems headed for destruction. The very urgency of the situation presses us toward short-term actions. Yet it is clear that much of our current dilemma derives from making decisions without regard for (or ability to anticipate) their consequences in the future. In this case, consequences in the future often means long-term consequences in the far future. Arne Naess called the article that provided the term *deep ecology* the "Deep, Long-range Ecology," and it has now become a commonplace among environmental activists that work for sustainability will need to be done with a long-range perspective.

The same dilemma arises when we think in more personal terms. When we think of our own futures, our hopes for ourselves, we are fully engaged in the present and the near future with its possibilities. When we think ahead to the fates of our descendants, however, we feel the anxiety that the current situation inspires. It is hard to hold both perspectives, the short-term and the long-term, the near and the far, ourselves and the larger world, together in our minds. Yet our concern for those who will come after us impels us to find a realistic hope for the future. What do we need to do in order to address the larger question of the future? What would make a long and livable future for all of the world possible? And how are we to think about our engagement with the process by which the future comes into being, when the future seems compounded of possibility and uncertainty and influenced by a whole variety of mental activities, from dreams and visions to desires and hopes to predictions and plans?

Leadership, Group Life, and the Transmission of Emotional Knowledge

Human beings have a natural tendency to be frightened of each other. This fear of each other is a fear of the unknown, both in the other and in ourselves. . . . To overcome this state of affairs, this fear, is the emotional task.
—Neville Symington, *Narcissism*

It is painful to be alone in a world of wounds, said Aldo Leopold. Leopold was thinking of the special awareness that ecological knowledge brings: one sees the wounds, while others seem to be unaware, oblivious, unconcerned. Hence one is alone. Is this so true? That the world is a world of wounds rings true, but how alone in it are we really? The brunt of what I have argued in this book is that our profound attachments—to people, places, and the natural world—entangle us with all that surrounds us (the "environment"). Perhaps our human group life impels our destructiveness and solidifies our obliviousness and apathy. But our group life also makes it possible for us to imagine a future in which healing is occurring. This future is certainly one in which we are not alone. As Thomas Berry would say, it is one in which we recognize again that we are not alone but rather part of the community, or communion, that is the earth and its creatures and life forms.

In his book *Leadership Without Easy Answers,* Ron Heifetz distinguishes between technical and adaptive challenges and between leadership and formal authority.[3] These distinctions, derived from Bion's group work, are illuminating in contemplating our situation in a world of wounds. In Heifetz's thinking, a technical challenge is one that we have already developed routines and procedures for dealing with. It requires no new social learning. An adaptive challenge, in contrast, is one that will require us as a society or species to learn and change, to be transformed. Formal authority is the ability to direct, orient, protect, to manage technical routines. The authority is entrusted to the person in the role by the others. Leadership, in contrast, is the capacity to mobilize activity in the service of the learning needed to meet the adaptive challenge.

What we are facing now in recognizing our loyalties to both people and places, the human and the natural world, and in learning to live in a nondestructive way, as part of the whole, is an adaptive challenge of overwhelming proportions. Bion, whose thinking gave rise to many of Heifetz's conceptions, pointed out that while we are able to pass on our technical knowledge by imitation, the transmission of emotional knowledge—what he calls the capacity for learning from experience—is incredibly difficult. If we are to meet the adaptive challenge of the "movement toward dissolution" Berry speaks of, we will need leadership—a capacity to mobilize learning from experience on the part of society as a whole.

In 1947, Bion delivered a paper called "Psychiatry at a Time of Crisis" to the Medical Section of the British Psychological Society. The time of crisis he referred to was not simply the horrors of the Second World War; it was the whole previous forty years, since the time before the First World War. That period had brought tremendous changes, both in the form of rapid technical advances and in the form of barbarism and overwhelming personal suffering. In his paper, Bion asserts that we have failed to grapple with the main problem of our current civilization—dealing with the underlying emotional tensions in human relationships.[4] He contrasts technical advances, which are easily communicated and can be learned through imitation, with emotional development, which is tremendously difficult to communicate. Indeed, he notes, "no method of communication of emotional development has yet been found which is not hopelessly limited in its field of influence."[5] Certainly, emotional knowledge— understood as development—cannot be acquired through imitation. When it comes to emotional and intellectual development, in fact, imitation is dangerous, because it gives the appearance of growth when none has occurred. As a consequence of this dilemma, when it comes to technical matters, the fate of societies is in the hands of the most gifted, whereas in the emotional domain it is in the hands of the least advanced.

In dealing with the problems of personal relationships, we cannot rely on technical reasoning, or even on abstract reasoning in isolation from feeling. We have to try to understand the emotional tensions in the social group. When emotional problems arise, however, Bion points out, people tend to retreat into further exploration of the possibilities of external regulation. "These repeated preoccupations with machinery," he writes, "being themselves nothing more than a by-product of the failure to grapple with the main problem, never achieve any higher level than the technicalities of police operations, no matter how magnificent the language in which they are clothed."[6]

Bion's comments, more than fifty years old now, seem particularly apposite today. The craving for technical solutions to problems of group emotionality is obvious not only in the social field—"providing jobs" by building prisons— but in our vain hopes of thereby dealing with increasing environmental disintegration and the destruction of life. The need for understanding the emotional tensions in the social group is more crucial than ever.

Morale and Psychiatric Disinheritance

Society has not yet been driven to seek treatment of its psychological disorders by psychological means because it has not achieved sufficient insight to appreciate the nature of its distress.
—Wilfred Bion, "Intra-Group Tensions in Therapy"

During and after the Second World War, Bion experimented with ways to help groups examine the emotional afflictions within the group. The result of his work with groups was twofold: heightened morale on the part of the group on the one hand, and intense resistance on the other. Bion published his reflections on this work in his book *Experiences in Groups*. As we have seen, Bion's idea is that humans are both individuals and herd animals with a need for group—that is, political—activity. We cannot get away from that need; we cannot escape into mechanical activity. The cooperative power of the thinking individuals gathered in the group allows the group to act on problems of reality. At the same time, the group attempting to do collaborative work always has to contend with the powerful emotions that unite its members in an unconscious "group consciousness" that Bion calls the *group mentality*.[7]

This unconscious group consciousness, Arthur Colman notes, expresses itself in archetypal imagery, dreams, visions, myth. In his book *Up from Scapegoating*, Colman describes a widespread developmental sequence of such myths in groups, from the matching myths of scapegoat and messiah, to the myth of the self-contained island, to a myth of interdependence like King Arthur's Round Table.[8] At its most unconscious, at the level of development that Colman talks about in terms of the scapegoat and the messiah, the group is most dangerous. Individuals with their moral perspectives are swallowed up in the group consciousness. The scapegoat and the messiah: think of the interlinking apocalyptic hopes and apocalyptic terrors that accompanied World War II, the death camps, and the development of nuclear weapons. Why did the development of nuclear weapons continue when, from a number of points of view, it was no longer necessary? Robert Wilson, the Los Alamos scientist quoted in the previous chapter, provides a good picture of a "group mentality." He did not act on his own values, nor did his colleagues. They were acting, he says, like "automatons"—that is, like human beings who had lost the capacity for individual thought and judgment.[9]

These examples make it clear that difficulties in transmitting emotional knowledge have disastrous social consequences. Again, as Bion said, we have to try to understand the emotional tensions in the social group. It may well be, however, that the technical and organizational structure of a particular society has created a situation of what Bion calls "psychiatric disinheritance" in which all but the "dominant minority" (he is following Toynbee here) are prevented from functioning as fully responsible citizens.[10] This would inhibit the development of leadership, which requires responsible followership. The ability to make good group decisions thus depends on the restoration of individuals in the social group to their psychiatric inheritance. When individuals, or whole classes of people, are psychiatrically disinherited, in Bion's term, they are unable to maintain their individuality and function responsibly—as fully capable citizens—within the society. In our current situation of urgency we will need our full powers of collaboration to make a future possible. To do that, we will need to understand ourselves in groups.

In 1940 Bion published a paper called "The War of Nerves" in a volume in which psychiatrists attempted to explore the psychological dimensions of war. Bion begins this paper by saying that in a war situation the enemy's aim is to make "psychological casualties" of all of us. We can become psychological casualties in two ways: first, by dissociating ourselves from our community (the group), and second, by being overwhelmed by anxieties—"unduly alarmed by the nightmare noises of others," in Bion's phrase.[11] In other words, we can be lost in the primitive terrors of the group mentality. In war, he writes elsewhere, "the enemy's object is so to terrify you that you cannot think clearly, while your object is to continue to think clearly no matter how adverse or frightening the situation."[12] If we think of the enemy as the forces, internal and external, tending toward what Lifton calls atrocity-producing situations, this certainly describes our challenge today.

"The War of Nerves" centers around the problem of morale—in other words, the problem of a group's capacity for effective collaboration. Specifically, it addresses the problem of what can be done to help English civilians avoid becoming psychological causalities during the anticipated air raids. Bion's advice is telling: people need to feel that there is a larger structure of trustworthy authority whose thoughtfulness and foresight are evident. And people need to feel that there is a way they can contribute so that they are not "condemned to a harassing passivity."[13]

In our current situation, with the future unknown but menacing, we too need to feel a sense of cohesion in which foresight is active. We also need to feel that we can contribute something of value so that we are not condemned to passivity. But at the same time Bion emphasizes that the problem we face is that of understanding ourselves emotionally in our communal life. We cannot achieve cohesion and foresight if we are swept away by the group mentality, and we cannot escape passivity through technical mimesis. We will have to proceed through the much more anxiety-provoking route of learning to work effectively in groups.

Binocular Vision

We have heard of a society for the Diffusion of Useful Knowledge. It is said that knowledge is power, and the like. Methinks there is equal need of a Society for the Diffusion of Useful Ignorance, what we will call Beautiful Knowledge, a knowledge useful in a higher sense; for what is most of our boasted so-called knowledge but a conceit that we know something, which robs us of the advantage of our actual ignorance?
—Henry David Thoreau, "Walking"

We can believe whatever we please, but that doesn't mean that the universe is going to suit itself to our particular beliefs or our particular capacities.
—Wilfred Bion, *Attention and Interpretation*

In his book *Experiences in Groups,* Bion talks about his frustration in a particular group therapy session. While all the individuals seemed to be trying hard to contribute, somehow nothing seemed to go anywhere, and nothing Bion, the psychiatrist in charge, could think of to say seemed helpful. But then he begins to wonder, "What is this group which is unsympathetic and hostile to our work?" In fact, he concludes, he has to assume that it consists of the very same people that are trying so hard to do the work. "I am reminded of looking through a microscope at an overthick section," he writes. "With one focus I see, not very clearly perhaps, but with sufficient distinctness, one picture. If I alter the focus very slightly I see another."[14] With this idea, Bion becomes alert to seeing evidence of resistance in the group as well as evidence of a sincere effort to work.

This attempt at *binocular vision,* as Bion calls it elsewhere, became the foundation of Bion's understanding of humans' participation in groups. As we have seen, people require group life—communal life, political life—for their full development. At the same time, the developed individual is uncomfortable with the unconscious, gregarious impulse that leads people in groups to feel at one with each other in a kind of unquestioned and unreflective irrational emotionality. What Bion calls the *sophisticated group* or the *work group*—the mentality in a group that uses collaboration and the differentiated contributions of individuals to think together about problems with which reality confronts them—

cannot exist in isolation from the *group mentality* that Bion also calls the *basic assumption group*. (As we saw earlier, the basic assumption group is not a separate group of people but rather the group acting as though it were in the grips of a highly emotionally charged fantasy.[15]) Since there is no work group without the emotional forces of such group fantasies flowing through it, and since every member of the group, no matter how alert and independent, will also be participating in the group fantasy as well, binocular vision is required in order to be aware of what is going on.

The notion of a binocular vision is reminiscent of the imaginative vision that William Blake advocated as an alternative to the "single vision" of Newton, Locke, and the mechanistic philosophers. It will prove useful in thinking about the question of our possible relationship to the future, for as Bion noted, our capacity to meet the adaptive challenges we face will depend on our ability to collaborate effectively as thinking individuals.

Bion left group work at the end of the 1940s, but toward the end of his life he became preoccupied with a concept closely akin to that of binocular vision. He called it *penetrating the caesura*.[16] By this he meant being able to retain one's normal awareness while also becoming aware of a state of mind radically different from it. Remember that Bion's use of the word *caesura* was inspired by Freud's phrase "the impressive caesura of birth." Perhaps, thought Bion, the divide marked by birth, the change from uterine to extrauterine life, which seems such an absolute divide that it becomes a complete mental barrier to our capacity to think, can be penetrated by thought. Perhaps other such seemingly absolute barriers could be penetrated by thought as well. Must we always, for instance, think of the sleeping state as simply an adjunct or aid to the waking state, with which we identify? How would the waking state be experienced from the point of view of the sleeping state? To put this notion of penetrating the caesura in its most essential form, how can we penetrate the caesura of "as we all know"? In other words, can we move between our knowledge and our ignorance? It is as though our belief that we know, and that only one side of this divide can be known, constitutes a filler that plugs up the holes in our awareness. Those holes, however, might otherwise become open passages to new awareness and ideas. They might, in short, be pathways to creativity. (And does not the process of pregnancy and birth precisely cross that divide, not to mention the divide between male and female?) In our current situation, it may become extremely important to penetrate the caesura of "as we all know." As Bion said, "It is a question of whether . . . the answers that are immediately comprehensible, that can be used to fill up the space of our ignorance, mislead us into extreme danger."[17]

Penetrating the Caesura of Language

> *We live surrounded by possibility, not merely by already existing things. In the prison of mere already-existing-things we could neither move nor breathe.*
> —Ernst Bloch, "Man as Possibility"

> *Language is the main instrument of man's refusal to accept the world as it is.*
> —George Steiner

Neil Evernden ends his wonderful and surprising book *The Natural Alien* with an epilogue in which he suggests that the problem with human beings is that they are by nature placeless. As they develop the abstracting capacity characteristic of language, they become able to lay claim to the whole of the world. In this they are quite different from other species, who have a niche that allows them both to be secure (for "instinct," Evernden suggests, read "faith") and to coexist with other creatures, whose different *Umwelten,* or experiential environments, mean that they are not in fact competing for the same "resources." From this point of view, says Evernden, animals are "embodied limits." They are the fleshly visibility of place: "One might say that the place is the species, for the place is more real and enduring than flesh."[18]

"The growing human being is born out of the mother's body into a second uterus in which he traverses the second half of his embryonic life: this is the *social uterus,*" writes Adolf Portmann.[19] Culture, the "social uterus" Portmann speaks of, may provide humans an approximation of the meaning that animals have by virtue of their limitations—their niche. But when culture does not provide meaning, it produces the features of an exotic species in human beings, so that they become, in Evernden's words, "a collection of pathological selves with little potential for the constituting of reciprocal relationships."[20]

As one of the most important vehicles of culture, language can participate in providing embodied limits, but it can also aid and abet the rootless and exotic potential of the human species. As the vehicle of abstraction, language categorizes and thereby in some sense removes us from direct experience of things.

But perhaps, suggests Evernden, there is still the possibility of "an internal challenge, an attempt to 'un-say' our present world" and "at least aspire to some cultural imitation of 'embodied limits.'"[21] Phenomenology, as the philosophical approach that attempts to step back from categories and experience things more directly, might then provide humans with a sense of something more like embodied limits, which would in turn leave room for the other creatures. "Phenomenology, with its explicit shunning of explanations and constructs," writes Evernden in a statement strikingly reminiscent of Bion's interest in penetrating the caesura of "as we all know," "might usefully be regarded as a means of un-plugging the gaps which we have filled with words and symbols, so as to focus on that which is humanly significant and let the rest slip past for the habitations of others."[22]

We cannot fully remove ourselves from our cultural perspective or from the perspective shaped within us by our language. But we can attempt to become aware of the phenomenological perspective of direct experience as well. In short, perhaps we can have a kind of binocular vision that would allow us to be aware of both our symbolizing capacity with its categorial frameworks and our more limited and embodied experience. In this way Evernden's presentation, seemingly so differently focused from Bion's, suggests a direction that is much the same.

Evernden's evocation of the direct experience phenomenology is intended to give access to, points up the complexity of the unspoken. The unspoken has much in common with the "group mentality" Bion talks about, a mentality that is often antagonistic to curiosity and openness but that also provides the emotional basis for what becomes experience articulated in language. At the same time, though, the unspoken is the realm of that of which we are—so far—ignorant, an unknown to which we might hope to open ourselves through the direct experience of things.

Bion advised the psychoanalyst to try to suspend memory, desire, and understanding when working with a patient. Only then, he said, would the analyst be truly open to what was evolving from moment to moment within that always-new experience.[23] His advice, so extraordinarily difficult to follow, is intended to open the analyst to what Evernden calls the direct experience of things and what Bion himself would call the evolution of "O," or reality, ultimately unknowable except as it evolves through our open experience. This openness to what evolves in the moment, which is the new, bears a resemblance to the Buddhist practice of mindfulness. It is what we need if we are to experience the future in its evolution. But it requires extraordinary courage.

Courage, Truth, and Mindfulness

To dare to be aware of the facts of the universe in which we are existing calls for courage.
—Wilfred Bion

If one does not look into the abyss, one is being wishful by simply not confronting the truth about our time. . . . On the other hand, it is imperative that one not get stuck in the abyss.
—Robert Jay Lifton

In the early 1940s, as an army psychiatrist, Bion, in collaboration with his colleague John Rickman, was appointed head of the "training wing" at what was referred to as "Shell-Shock Hospital"—Northfield Hospital, near Birmingham, England, where war-traumatized military personnel were sent for rehabilitation. Here Bion tried some radical ways of working. Although what was later known as "the Northfield Experiment" was short-lived, it became famous. It was here that Bion's understanding of group work and its relation to morale and to leadership was really developed.[24]

Bion tells how on first arriving at Northfield he went to look around and was shocked to see strapping officers waving feather dusters in the occupational therapy unit, and a man shamefacedly carrying a papier-mâché figure while his serviceable gun lay disabled nearby. This was the community in full flight from the enemy, he decided—the enemy being neurosis: "It was essential first to find out what was the ailment afflicting the community, as opposed to the individuals composing it, and next to give the community a common aim. In general all psychiatric hospitals have the same ailment and the same common aim—to escape from the batterings of neurotic disorder. Unfortunately the attempt to get this relief is nearly always by futile means—retreat."[25]

To deal with the common enemy, neurosis, leadership was needed. Bion conceived leadership in terms of a good military officer: "I became convinced that what was required was the sort of discipline achieved in a theatre of war by an experienced officer in command of a rather scalliwag battalion." What Bion means by "an experienced officer" is this: "An officer who, being experienced, knows some of his own failings, respects the integrity of his men, and is not afraid of either their goodwill or their hostility."[26] This is the officer, or leader, who is both in touch with his men and concerned for the truth—someone, in other words, who has the courage "to dare to face the facts of the universe in which we live" and to work for the binocular vision that will reveal both individual interests and group solidarity, both rationality and one's "own failings."

At Northfield Bion started the men working in groups that had a genuine purpose. He put them into a situation of collaboration on real tasks—tasks relevant to their lives on the wing—rather than collusion in flight into neurosis. The morale of the group improved immeasurably. Here, then, the officer plays the role of a courageous leader against the psychological onslaughts of what Bion calls "neurosis." It is the leader who prompts society to take action on its psychological afflictions rather than conceiving them as confined to individuals. Morale, as the capacity of the collective to attend to real work, to the genuine task at hand, is the social equivalent of mental health.

The courage to face the facts of the universe in which we live can also be conceived as a form of curiosity. The concern for truth Bion talks about implies an openness that we might call curiosity. While curiosity can take the form of an invasive prying, it can also be based in concern. In that case we can think of it as a hunger for the truth. Lies, in contrast, would serve to obscure the opening through which curiosity might find the truth. How courageous our curiosity must be in the current situation is indicated by how unusual it is for anyone to be curious either about the current state of the environment or about our possible future. It is surprising, that is, as I noted earlier, to find curiosity accompanying our concern for the environment. Concerned curiosity might, however, open the space to notice connections we had not seen before, to notice details of the current situation that might lead down unusual paths.

Where curiosity is a manifestation of concern, it indicates the presence of an individual state of mind within someone who is also part of the group. In other words, it is a correlate of binocular vision. The space for thought opened by binocular vision creates the possibility of further change, further transformation of the existing order. Bion referred to such change as *catastrophic change*,[27] in acknowledgment both of the tremendous upheaval it represents and the terror with which it is greeted. It raises the specter of violence to the existing order, internal and external, and in turn can evoke violent and aggressive resistance in all who might be affected by it. "The search for truth can be limited both by our lack of intelligence or wisdom and by our emotional inheritance," Bion notes. In fact, he adds, "The fear of knowing the truth can be so powerful that the doses of truth are lethal."[28] Thus nothing so threatens and enrages the group mentality as the presence of a thinking individual within the group. "It is those individuals who try to . . . maintain some semblance of individuality [in the large group] who are most attacked," he writes; "it is as if, in the large group, there were a general envy of people who maintain their sanity and individuality."[29]

Bion and Rickman's Northfield Experiment was terminated after six weeks, and Bion and Rickman were transferred. Though there are conflicting stories about what might have led to the termination, it seems clear that the experiment had aroused opposition within the hospital and military hierarchy.[30] In his autobiography, *All My Sins Remembered*, Bion records some of his inner dialogue about this outcome of the experiment:

> Pearce told Rees, and Rees did not need any telling, that it was dangerous to have people as "intelligent" as Rickman and me around because either we should blow up the Military Training Scheme (and the whole of Army psychiatry) or one of the Big Guns would fire at us, and they— Pearce and Rees—would get hit.
> *So: what happened?*
> Rees had us posted off to where we could do no harm.[31]

While it may well be the case that Bion failed to anticipate the resistance his experiment would arouse among the hospital and military authorities, it is also clear that he understood the experiment to be opening a space for thought in the group in a way that conflicted with the rigid distinctions of a structure in which some of the participants—the patients—were "psychiatrically disinherited." "So I shall never know what would have happened," his inner dialogue continues, "if privates in the Training Wing had caught the habit of asking questions or having opinions of their own which they might be able to hear in the silence of the sleep time when their fancies could become free to roam like the wind."[32]

Generosity and the Labor of Gratitude

When I heard her sing, I felt I was in the presence of a great leader.
—Interviewee in a film on the Egyptian singer Uum Kulthum

[Followers] must be, and yet seldom are, capable of leadership and capable of something more than imitation of the original thinker.
—Wilfred Bion, "Psychiatry at a Time of Crisis"

Leadership requires courage. The leader's courage in "facing the facts of the universe in which we live" allows others—the "followers"—the opportunity to face them as well. Generosity is an equally important, though less frequently acknowledged, characteristic of leadership. The question of how followers become collaborators in a "work group" points us, perhaps surprisingly, to the role of generosity.

Lewis Hyde's ideas about the gift, which we met earlier, are helpful in understanding this link between generosity and followership. Hyde distinguishes between the work of art as art, which is freely given, and the work of art as commodity, which is exchanged. In some essential sense, the artist "gives" the world the work of art, even though he or she also earns a living by making art; similarly, the "gifted" artist's talent is given to him or her. As we have seen, the artist earns the right to the gift of talent through what Hyde calls the *labor of gratitude.*

Gifts like the artist's talent fall under the general category of what Hyde calls *threshold gifts.* Such gifts are used to mark a rite of passage at one of life's stages. The prototypical threshold gift is the food and alms that are handed to the poor across the coffin by the family of the person who has died. The artist's talent is a special kind of threshold gift, a transformational gift, in which the gift itself is the agent of transformation in the recipient. A transformational gift, as we have seen, does its work through the labor of gratitude.[33] Followers too receive threshold gifts through the generosity of the leader, through which they become followers capable of leadership rather than imitation.

Let us look again at Hyde's retelling of the story of the shoemaker and the elves in this context. The elves' capacity to make beautiful shoes is the "gift" or "talent" that the shoemaker acquires through the labor of gratitude. He learns how to make shoes like the elves do, and the end result of his labor is that he can make the elves his gift of shoes, thus closing the circle of gift exchange. The shoemaker now "owns" his expertise in shoemaking. The end result of the labor of gratitude, says Hyde, is similarity with the gift. The shoemaker becomes able to do what the elves could do.[34]

Something this tale does not point up, however, is that similarity with the gift does not mean identity with the gift. Protégés who ape the master have not succeeded in owning their gift, and the master will not be pleased with them. For the gift to be acquired is the gift of spontaneous creativity (supported, of course, by hard work), and imitation is not creativity. As Bion notes, "[Followers] must be, and yet seldom are, . . . capable of something more than imitation of the original thinker."[35] What the followers do through their exposure to the leader's generosity is to assimilate a capacity for leadership. Working to make good use of what they receive from the leader is their labor of gratitude. In what Bion calls a work group, this process of receiving and using the leader's generous gifts of binocular vision takes the form of active collaboration on the task at hand. In the fantasy-based group mentality that Bion calls a *dependency group,* in contrast, the leader is imagined to be withholding rather than generous with these gifts. The leader is resented and envied, and the leader's generosity is neither acknowledged nor accepted.

Neville Symington's theory of narcissism is also helpful in understanding this relationship between generosity and followership on the one hand and receptivity and creativity on the other. As the reader will remember, Symington understands narcissism as the (unconscious) decision to reject the Lifegiver, the source that gives life to us.[36] Initially the Lifegiver is an external source. If we accept it, we internalize it, and it becomes a source of what Symington calls "initiatory," or creative action. Of course no one can fully reject the gifts of life; that would mean death. But the part of oneself that rejects the Lifegiver and imagines itself to be autonomous is the part that is narcissistic.

The narcissist in Symington's understanding is not enamored of his or her self. In fact, there is no deep self there for the narcissist to love, since narcissism is precisely a rejection of that which allows a deep, authentic self to develop—the Lifegiver. Lacking a deep self, the narcissist can only proceed by a kind of blind imitation. Here we are in the realm of Bion's "monkey tricks," which are based on imitation. In contrast, the "transmission of emotional knowledge" that Bion considers so important, and so difficult, has to be the result of a transformative gift, which is then earned in turn through the labor of gratitude.

Symington's Lifegiver is an internal source of generosity. It makes initiatory action possible. In a definition that resonates with Heifetz's conception of leadership as facilitating social learning, Symington describes this initiatory activity as a "creative fashioning of the social environment." He gives a lovely example of such creative fashioning. The example is taken from the autobiography of Graham Greene, who is describing the British art critic and writer on aesthetics Herbert Read:

Certainly my meeting with Herbert Read was an important event in my life. He was the most gentle man I have ever known, but it was a gentleness which had been tested in the worst experiences of his generation. The young officer, who gained the Military Cross and a DSO in action on the Western Front, had carried with him to all that mud and death Robert Bridge's anthology *The Spirit of Man,* Plato's *Republic* and *Don Quixote*. Nothing had changed in him. It was the same man twenty years later who could come into a room full of people and you wouldn't notice his coming—you noticed only that the whole atmosphere of a discussion had quietly altered, that even the relations of one guest with another had changed. No one any longer would be talking for effect, and when you looked round for an explanation there he was—complete honesty born of complete experience had entered the room and unobtrusively taken a chair.[37]

When followers are able to make use of the leader's generosity to develop their own capacities, including their own capacities for binocular vision, the transmission of emotional knowledge has taken place. The labor of gratitude that acknowledges both one's dependency, on the leader and as part of the group, and one's own separateness from the group as an individual, has been completed. The Lifegiver has becomes an internal source of generosity that makes the person's own creative leadership possible.

The Dream and the Spell

> *This arid and desperate moment of time wherein we live . . .*
> —Laurens van der Post, *The Voice of the Thunder*

> *We can counter one entrancement only with another, a counterentrancement. Only thus can we evoke the vision as well as the psychic energies needed to enable the Earth community to enter successfully upon its next great creative phase.*
> —Thomas Berry, "The Ecozoic Era"

Thomas Berry is emphatic that we must dare to face the facts of the universe in which we live. This era is coming to an end, he insists. The present system, the industrial world, cannot continue. It is moving toward dissolution. The Cenozoic era in which we have been living, an era marked by a magnificent flowering of species, is waning and will not return. But Berry also looks to a possible future beyond this dissolution. Thinking about the next generation sharpens the question of the future. What will the next generation face? Certainly it will face the facts of dissolution. Can we provide it with inspiration, with a link to the possible as well? "I hope we will be able to guide and inspire our next generation as they attempt to shape the future," he comments. The alternative is bleak: "Otherwise they will simply survive with all their resentments amid the destroyed infrastructures of the industrial world and the ruins of the natural world itself. The challenge itself is already predetermined. There is no way for the new generation to escape this confrontation."[38]

How is a new era created? Possibility becomes reality, Berry tells us, by virtue of its existence in the collective mind as a dream; the dream is infused with emotional energy that provides the motivation for its realization. The dream that gave rise to the industrial world is played out. To capture and realize other possibilities means that a different dream—Berry's "dream of the earth"—must take shape in the collective mind. Berry calls this a *counterentrancement*: a new dream that will replace the old dream that has become destructive.[39]

Berry's term *dream* emphasizes the longing and hoping that entered into even the destructive dream in which we are currently caught up. His term *entrancement* brings to mind the other side as well: the fact that we have surrendered to some force that we feel to be other than ourselves. We are under its spell.

In *All My Sins Remembered,* Bion tells a story about being under such a spell. Bion's first wife had died in childbirth during the Second World War. The incident Bion recounts took place in England after the war, while Bion was raising his young daughter as a single parent and working as an analyst. He writes:

> Patients came and paid. My fees, though small, were wealth to me and I could even afford—financially—meals that were bad for my health. So I could escape from the pains of indigence into the pleasures of gluttony.
>
> Yet now I felt, as never before, numbed and insensitive. That something was wrong, must be wrong, was brought home to me one week-end when I was sitting on the lawn near the house and the baby was crawling near a flower bed on the opposite side of the lawn. She began to call out to me; she wanted me to come to her.

I remained sitting. She now made to crawl towards me. But she called to me as if expecting me to come to fetch her.

I remained sitting.

She continued to call and now her calls became distressful.

I remained sitting.

I watched her continue on the painful journey across the vast expanse, as it must have appeared to her, that separated her from her Daddy.

I remained sitting but felt bitter, angry, resentful. Why did she do this to me? Not quite audible was the question, "Why do you do this to her?"

The nurse could not stand it and got up to fetch her. "No," I said, "let her crawl. It won't do her any harm." We watched the child crawl painfully. She was weeping bitterly now but sticking stoutly to her attempt to cover the distance.

I felt as if I were gripped in a vise. No. I would *not* go. At last the nurse, having glanced at me with astonishment, got up, ignoring my prohibition, and fetched her. The spell snapped. I was released. The baby had stopped weeping and was being comforted by maternal arms. But I, I had lost my child.

I hope there is no future life. . . .

It was a shock, a searing shock, to find such depth of cruelty in myself.[40]

Bion's story conveys well the sense of being under a spell—what Berry would call an entrancement—a spell that prevents obvious action from being taken. It makes him "remain sitting." Under the spell, he watches powerlessly while suffering occurs. But this inaction is cruel; it results in further suffering. Once the spell is broken, he sees the depth of cruelty in himself and experiences terrible guilt. These realizations have a devastating impact on his sense of a future: "I hope there is no future life," he says, meaning that his guilt, his "sin," is such that the memory of it—his sin—is now and would be in the future unbearably painful to him. But he has also "lost his child." The child, his daughter, also represents life going on into the future. In this way too, future life has been destroyed; Bion feels he has lost the ability to relate to what will come. Being committed to trying to bear pain for the sake of truth, however, Bion writes this story in his autobiography and calls the book *All My Sins Remembered.* This is his effort to come to terms with the dissociation and cruelty involved in the spell he was under, and his guilt.

We can think of such cruel entrancements as failed responses to loss. Bion fell under this spell after a terrible loss and strain—the loss of his wife, the destruction of the family context. Loss not fully faced and mourned can lead one to fall under the spell of numbness and cruelty. It can lead to guilt and the wish to blot out the future. We can also think of such spells as a form of narcissism, a revulsion against our dependency on one another and on the earth. Perhaps Bion's response to his daughter was a momentary revulsion against her dependency on him—evoking the terrible memory of his first wife's dependency on him and his on her, a memory linked to her death and his feeling that he abandoned her in her moment of need. Similarly, the industrialism that for Berry has become a destructive entrancement can be seen as a rage against the earth and a delusory attempt at control, an effort not to need or be dependent on the natural world, which remains beyond our control.

In this revulsion against dependency, some part of ourselves asserts its utter autonomy and holds the more dependent part hostage, as it were, threatening and silencing it. (This is the menacing willfulness we see in some antienvironmental pronouncements.) Some theorists refer to this as *destructive narcissism,* and Symington has shown us how life denying narcissism is. Dismantling this destructive narcissism entails experiencing despair, both despair over what we have deprived ourselves of, and despair over what we have wrought in our delusion. This despair is felt as apocalyptic. As Michael Ortiz Hill conceptualizes it in his book *Dreaming the End of the World,* "Seeking refuge in an apocalyptic era is the flipside of recognizing that, ultimately, no refuge exists."[41] This is the despair that made Bion wish there were no future life.

It is important to acknowledge how reluctant we are to settle our questions, to accept the obvious and deal with the consequences of facing the truth. As Berry says, we are reluctant to think that what we are doing is inherently incompatible with the workings of the larger world. Think how much information we have put forward and how many times we have written things that point so compellingly to the same conclusions, yet we keep debating the questions over and over again. It is as though our inconclusiveness were itself a spell, keeping us paralyzed in a suspended, timeless state.

Why this temporizing? Why is it so difficult to accept that something really cannot be saved, that a change must occur? To awaken from the spell of timelessness is to see that time is real, bringing change, bringing death as well as life. To accept that something is really over, really finished, means to mourn, to grieve. We need to mourn for this dying way of life that we have lived. As a species it has not been with us all that long, but it has lasted throughout our lifetimes. Let us call it, with Berry, the *petroleum era*. It is difficult to mourn this passing way of life not only because we are so attached to it and to so many parts of it, but also because it has been, in some sense, our *whole* life. To think that it is, or will shortly be, over, or that it needs to end is to envision a kind of death of ourselves and our way of living. And in addition, to accept that this way of life is over is to face its destructiveness and our involvement in that destructiveness. How can one mourn for something destructive? How does one mourn when one is guilty?

To endure the change that is occurring, we will need to suffer not only the pain of privation but also the pain of mourning and the pain of guilt, as well as the pain of changing our understanding of who we are. We will suffer loss of pride in our achievements, of our status and prestige, as well as the loss of comforts and of established systems. We will need a new vision, a new dream, as Berry says, to sustain us in this transition. "The grandeur of the possibilities ahead of us," he writes, "must in some manner be experienced in anticipation. Otherwise we will not have the psychic energy to endure the pain of the required transformation."[42]

Burying the Dead

Who brought the dead to the rear?
When were the dead mourned?
What was the level of trust, safety, and social cohesiveness in the rear during mourning?
Who wept for the dead, and how were tears valued?
Who washed and prepared the dead for cremation/burial and shipment home?
—Jonathan Shay, *Achilles in Vietnam*

I think that planetary, ecological disaster is a reality. But the popular imagination is in error. It is not something that may happen. We have been in the midst of it for the last century.
—Paul Shepard

Certainly it is possible to learn from experiences that are painful and destructive as well as from experiences that are nourishing. To learn from the experience of what we have actually done—of our collective actions in the petroleum era—will mean shifting to a state of mind in which we take cognizance of the consequences of our actions. This shift will require a tremendous amount of mental containing capacity, of mental space and strength. Hopefully this capacity will be fostered by cultural forms, as C. Fred Alford proposes.[43] For what we have done is horrible. We have committed acts of murder and violence almost without end, and we will need to go through a massive process of mourning. The reality of extinction, the reality of habitat loss, the reality of lives lived in pathological states of frenzy, the passing of the beautiful age of flowering— these are things we have barely begun to acknowledge, much less come to terms with. Somehow we will have to undertake to bury the dead, with all that implies. As Laurens van der Post writes concerning our destruction of the Bushmen's culture in the desert of Southern Africa, we need to remember the Great Memory that remembers us:

> Somewhere, waiting to be unlocked in that memory, I am certain, is a pattern of life that will transcend and transfigure this arid and desperate moment of time wherein we live, and the first move towards that unlocking, I believe, is to absorb the horror of the story of what we have all done to the first man of Africa and the horror of what we are doing now to the plants and the animals that for one thousand million years at least were safe and multiplying in his and the desert's keeping.[44]

Mourning does not mean "forgetting and going on." Instead, it means remembering. It means fully accepting the reality of death and loss. Mourning entails the loss of a sense of omnipotence, because it means that death is real, that things are beyond our power. It means relinquishing certain hopes, in particular the hope for a life without death and loss. It means, that is, as Harold Searles says, integrating the relinquishment of some of our hopes with feelings of disappointment, discouragement, grief, loss, and rage.[45] In the end, though, mourning means "learning from experience" in that something of the experience is taken inside and becomes part of who we now are.

As Shay points out, mourning is facilitated by, and probably requires, cultural forms that allow grief to be communalized. "Any blow in life will have longer-lasting and more serious consequences if there is no opportunity to communalize it," he notes.[46] Part of Shay's book *Achilles in Vietnam* is devoted to the contrast between the opportunities for full mourning among the Greeks as described in the *Iliad* and the things that made full mourning impossible in

Vietnam. Who buried the dead in Vietnam? he asks. It was not those who knew and loved the dead. When were they mourned? They could not be mourned; there was no respite from danger. And so on.[47] And what opportunity do we have to mourn the forests clear-cut behind "beauty strips" that hide the devastation from the highway? Are the tears we shed for lost species valued by our communities? Who buries the dead animals who lie by the side of the road?

The consequences of a failure to mourn on the scale of the whole community are repression, apathy, deadness, and lethargy. As Bion notes in "The War of Nerves," failure to bury the dead is the sign of a serious deterioration in morale.[48] It also confirms and perpetuates that loss of morale. What are the effects of our collective failure to mourn the loss of childhood places, the loss of species, the loss of opportunities to eat fresh-grown food? Think of how much mourning remains blocked by the facile opposition of caring about people and caring about the nonhuman.

Frustrated rage is the residue of all this unmourned loss. "The emergence of rage out of intense grief may be a human universal," says Shay. "Long-term obstruction of grief and failure to communalize grief can imprison a person in endless swinging between rage and emotional deadness as a permanent way of being in the world."[49] We know this well in the case of traumatized war veterans. We saw it in Bion's cruelty to his young daughter after the loss of his wife and the stresses of the war period. I think it boils under the surface of our society as a whole.

Hope and Despair

> *If one only stresses the genocide and its inevitability, one is contributing to the massive destruction or the apocalyptic end result. If one probes the dimensions of hope, one is contributing to an alternative possibility.*
> —Robert Jay Lifton

> *The realm of hope seems to us a last repository of such innate goodness as human beings possess. One of the harshest maturational tasks the individual must accomplish, to become truly adult, is to realize and accept that his hope is "impure."*
> —Harold Searles, "The Development of Mature Hope"

Surely hope is necessary for a future to be possible. Life in its creativity unfolds toward what is new and as yet unknown. Hope, as an orientation to possibility, is an emotional expression of the ongoing creative process of life. Despair, in contrast, feels like an internal process of destruction, something that attacks the very life in us and propels us toward death rather than toward a future.

Optimism and hope are not the same. Optimism is both more detached and more deluded than hope. In optimism, we predict that the future will be good. There is a semblance of detachment that contrasts with the intimacy of hope. Does this detachment mean that optimism is rational? Surely not; optimism implies a willfully positive prediction, determined to see a continuing increase in whatever one is optimistic about. As Christopher Lasch points out, "Progressive optimism rests, at bottom, in a denial of the natural limits on human power and freedom, and it cannot survive for very long in a world in which an awareness of those limits has become inescapable."[50]

If optimism crumbles under examination, what of hope and despair? Hope, as Harold Searles points out, is a multifarious thing. We generally harbor many hopes, often contradictory hopes, and often self-serving and selfish hopes, perhaps vengeful hopes, hopes for revenge. It is not until hope is tempered with concern and with disappointment that it becomes what he calls "mature hope."[51]

Bion calls one of his fantasy-based group mentalities the "pairing" group. The pairing group exemplifies the problematic aspects of hope.[52] The term refers to a group in which the interchange between two members seems full of promise, and the other group members watch in hopeful fascination to see what the outcome of their interchange will be. The problem is that in order to maintain the assumption that this interchange will generate something unknown and wonderful, it is essential that the hope not be realized, that the new thing generated never come to be. For any actual thing generated will be experienced as falling short of the ideal, of what was hoped for. Hope here means permanent deferral of realization. To be hopeful means to live in the future while ignoring the present.

Perhaps it is disappointment that the pairing group needs to face in order to attain binocular vision, perhaps especially disappointment in the idea of infinite progress. Evernden points out that we may find a realistic hope in what we currently scorn: other animals. "Far from being tragic cripples, unable to aspire to the heights open to humans," he writes, "those 'other nations' [the animals] constitute a source of hope, for they are portentous evidence of the only successful experiment in living ever conducted on this planet."[53]

Is Evernden suggesting that the animals constitute a source of hope precisely because they do not aspire? Consider the difference between hope and desire. Hope does without, Harold Boris points out, putting its expectations into the future, whereas desire looks for what is available in the present.[54] We often think that it is our desires in the present that are leading to environmental destruction. It may equally well be our insatiable aspirations for infinite "progress." Are we capable not only of hoping for the good but of tolerating the unexpected when it appears in our midst and inhabits our present with us? Paul Shepard, asked whether he had hope, responded somewhat cynically. "Of course I have hope," he said. "Why not, it's cheap and available. It is also the last resort."[55]

We often think of despair as the opposite of, or perhaps the alternative to, hope. But despair may consist not so much of hopelessness as of an unwillingness or inability to desire. If the future is foreclosed, desire becomes meaningless. Shay points out that one of the consequences of combat trauma is that veterans cannot participate in democratic decision making because they have no capacity to conceive of desire: "Democracy embodies safe struggle over the shape and implementation of a future. An unhealed combat veteran cannot think in terms of a future. . . . Combat taught the survivor of prolonged combat not to imagine a future or to want anything. Prior to seeing the point of one's voluntary participation in a social process, one must feel that it is safe to want something."[56]

Despair has its hopeful aspects. "Despair ushers in the death experience and is at the same time the requirement for resurrection," writes James Hillman, pointing to the pivotal role of despair in profound psychological transformation.[57] Neville Symington notes the rageful despair that the person emerging from a narcissistic delusion of self-sufficiency must experience.[58] We can think of this psychological transformation as the renunciation of a blind optimism and an acceptance of dependency. "Transformation begins at [the] point where there is no hope," Hillman writes. He elaborates: "Despair produces the cry for salvation, for which hope would be too optimistic, too confident. . . . My only certainty is my suffering, which I ask to be taken from me by dying. An animal awareness of suffering, and full identification with it, becomes the humiliating ground of transformation."[59]

This acceptance of suffering and dependency makes possible an experience of the present as the ground from which the future emerges. "Life as it was before," Hillman continues, "the status quo ante, died when despair was born. There is only the moment as it is—the seed of whatever might come—if one can wait. The waiting is all."[60] In the reversal of perspective that despair as acceptance of dependency implies, we may see disaster as a gift from what Laurens van der Post calls the Great Memory: "Never has it been so important and urgent, if human life on earth is not to fail the purpose which created it and gives it meaning, to begin again its ancient quest of seeking to remember what remembers us. It is because of my own brief experience of the power and the glory of this memory made manifest in the desert that I am certain, in its own time and after its own fashion, it will not allow us to fail, even if that means inflicting on us, as it has so often had to do in the past, the disaster we need to heed."[61]

The Generations

The possibility of death without rebirth is becoming our myth; it will remain our myth for the foreseeable future. It is not necessarily a story of a sacrifice for a new order. It may find its source in the ultimate act of species death, or it may be far more positive, a new process for living together.
—Arthur Colman, *Up from Scapegoating*

Many years ago, Margaret Mead wrote about a shift in the locus of social leadership that took place in the transition from traditional cultures to contemporary Western industrialized cultures.[62] In traditional cultures the elderly were looked to as the guardians of wisdom. In contemporary industrialized culture it is the young who are looked to. The young remind us that a future is emerging. They represent the possibility that is the hallmark of future time. Without a sense of time emerging, of possibility cradled in experience or shaped by wisdom, however, the young disappear into the void of timelessness. This is the problem of the contemporary youth orientation Mead described.

As Searles points out, the older generation may revenge itself on the younger by abandoning it to the havoc it has wreaked. Sending the young off to war is the epitome of this revenge. Abdicating the work of passing on wisdom is a more passive form of revenge on the part of the older generation. One of the forms generous leadership takes is the gift of wisdom from the older generation to the younger. In his book *The Force of Character,* Hillman talks about the faces of the elderly as the bearers of culture.[63] History is written in these faces.

They show the younger generations that grief and loss survived help to form the face that has endured. A culture that is passed on through the generations, such faces tell us, is a culture that is capable of mourning. A society that tries to deny aging and to hide the force of character that emerges in aging faces creates an illusion of timelessness that is also the illusion of deathlessness—and thus an obstruction of mourning.

The continuity of life through time is what makes one feel held and sheltered rather than isolated and apart. To know that leadership passes through the continuity of generations is to feel part of the human community. It is to know that one's concerns survive the passage from one generation to the next—that in fact they live by being passed from one generation to the next. Someone I knew grew up in a small town. She never heard anyone talk about her concerns about environmental destruction. Think of the isolation and confusion she felt, imagining—rightly or wrongly—that no one else shared her concerns, that no one else thought the thoughts she did. Think what it would have meant to her to discover a whole group of other people working on the basis of those concerns and those ideas. Later she did discover and join such a concerned group, and it meant a great deal to her to be part of it. People in that group, however, were generally around the same age. But then she found herself in an activity—a nonviolent protest at a hazardous waste burning site—in which a number of older women were taking part, while at the same time young mothers were bringing their children to greet the protesters. Finding herself part of the continuity of generations was a deeply moving experience for her. The possible future needs that support.

The Far Future and the Abyss of Infinitude

Is it by its indefiniteness it shadows forth the heartless voids and immensities of the universe, and thus stabs us from behind with the thought of annihilation when beholding the Milky Way?
—Herman Melville, *Moby Dick*

I think it may be erroneous to assume that because there is a past that seems to bear a certain resemblance to the present the present therefore bears a resemblance to the future, which can be described in terms of the past.
—Wilfred Bion, *Cogitations*

Elliot Jaques advances the notion that people's level of maturity is indicated by the time span they can encompass within a plan that will require evolution and thinking during the whole of its span.[64] Presumably, he says, people's time span gets longer as they develop and mature. Jaques's idea points up the tremendous disparity between the relatively short-term plans most of us make and the much longer time spans involved in some forms of environmental change. And what of geological time and the time required for the evolution of species, clearly relevant to our environmental concerns? What is involved in such a sense of time, or the lack of such a sense? What does one need psychologically to grasp not simply that time passes and that reality evolves within it, but also the far future into which the far-reaching consequences of our actions extend?

One difficulty in holding a sense of time that stretches into the far future is that the future, as possibility, is infinite. When Ernst Bloch tells us that "what is usually called reality is surrounded by a gigantic ocean of objectively real possibility," or that reality "only appears to be simple and solid" and is "a category which is exposed to flexibility and which is obligated to change,"[65] he draws attention to the possibility of making things different. But in doing so he also points to the unbounded, oceanic quality of possibility. Infinity can feel terrifying and timeless. The immensity and unknownness of future time can evoke an anxiety that feels urgent and is interpreted as the threat of imminent catastrophe. Paradoxically, in this urgent state of mind, time seems constricted and even nonexistent. The future is now and it is urgent, but there is no time to attend to it. The sense of urgency demands immediate action. Such urgency inspires in turn a wish to control, and we see all the responses such desperate anxiety gives rise to, from fantasies of omnipotence to manic activity to clinging to various forms of sensory stimulation.

It may be true that the future consists of a gigantic ocean of objectively real possibility. Still, when we dare to face reality, we see that as embodied beings we have limits. We live in time and cannot know the future; we are dependent creatures rather than omnipotent creatures. We must make choices if we are to live at all, and choices are limiting. Limits in turn are both reassuring and anxiety provoking, because they involve recognizing uncertainty and vulnerability. How do we hold both the infinity of the far future and the reality of choice and limitations in our minds at the same time? We need, I think, to have binocular vision here as well—to be aware both of the abyss of infinitude, doubt, uncertainty, and possible disaster *and* of limits and choice, which particularize us into a time that unfolds into the future.

Chilean psychoanalyst Ignacio Matte Blanco developed an understanding of the infinite and the particular that helps flesh out this binocular view. He called his theory *bi-logic*.[66] Matte Blanco distinguished between the conscious and the unconscious in terms of the difference between symmetry (sameness) and asymmetry (difference, which could include similarity). Asymmetry, which requires attention and thinking, localizes and particularizes: this man is this individual man, here and now. Symmetry assimilates and aggregates. It infinitizes, so to speak: this man is all men, in fact, is all human relationships. Both symmetry and asymmetry are necessary to mental life, but they are experienced very differently. Hence the term *bi-logic*.

Emotional depth involves symmetry. Eric Rayner, an expositor of Matte Blanco's ideas, provides an example: "Being in love"—a highly emotional state of mind—is an infinite rather than a particularized state of mind. I am in love, not only for the moment and in this place, but for all time into the future and wherever we may be.[67] Bion's group mentality too is very much a symmetrical matter. We are all the same, the group is one group (perhaps with a split-off opposition group), we are good, the enemy is bad, we are all helpless children and the leader knows all, and so forth.

This quality of infiniteness is an important part of elation—the sense of power and potentiality. But infiniteness and elation alone, unreconciled with particularity, are the qualities of mania and omnipotence. All things are possible, and I can do anything. As Rayner points out, mania and infiniteness depart from particularization and localization and are thus quite abstract. No one thing has any value; there is no focus.

Think of this in terms of our current environmental and social predicaments: capitalist-oriented globalization and "economic development" are based on abstract and infinite notions of commodity production, proceduralization, and standardization, notions that obliterate the particularity of place and local context. Such processes effectively "psychiatrically disinherit" not only the individuals who live in particular places all over the globe but also all other living beings. (This notion of the abstraction and infinitization of global capitalism is of course an extension of something Karl Marx perceived 150 years ago. Money as the medium of exchange, he said, is the abstraction that substitutes exchange value for use value.) To deal with problems in external reality, however, we require localization of attention as well as the broad categories that identify general issues and large-scale processes. This is why local control, sense of place, and collaborative decision making have become such prominent themes in our efforts to deal with the environmental and social crisis. They represent efforts to turn attention back toward particulars and away from the mania of growth, expansion, and abstraction. This emphasis on the particular is linked in turn with an acknowledgment of limits, another aspect of the "asymmetrical" side of bi-logic. As the term *bi-logic* reminds us, however, we must attend to both levels of scale, the global and the local.[68]

Springs of Living Water: Generativity and the Unborn

> *Though in many of its aspects the visible world seems formed in love, the invisible spheres were formed in fright.*
> —Herman Melville, *Moby Dick*

As a state of mind, infinitude is unborn and timeless. Reality is not yet real, and we are not located finitely in time and space. We have seen the anxiety the unborn state generates and the way it conduces to mania and omnipotence. Not yet being born—that is, not yet having become a separate and limited being—renders one unable to act, because action implies separateness. The actions taken, perhaps manically, by those whose mental state is at least largely unborn, do not reflect an orientation to reality.

But this unborn state has tremendous psychological importance. Precisely because it is outside of time and in touch with the infinitude of the evolving and unknown future, it makes the new possible. Mental—as opposed to physical—growth probably requires this contact with an infiniteness beyond our limits. Hence Bion's alarming statement, "Mental evolution, or growth, is catastrophic and timeless."[69] When we strive for moments of transcendence, as in meditation, we are hoping for a return to that ground of being so that we may start over, newly grounded, and go on from there. The "death experience" Hillman talks about is the equivalent—a chance to make contact with the infinitude of possibility and to reconstitute oneself. Surely this is what William Blake was referring to when he said that the hours of folly were measured by the clock, but the hours of wisdom no clock could measure. Where else but from this timeless and unborn state could new birth take place?

New birth, however, is birth into time and space. It requires a safe place, what Bion calls a "container" and others call a "holding environment," a capacious but bounded mental space, a place where a larger mental capacity is available to help digest the new experience. Perhaps it is the mother's capacity or one's own larger ego that can tolerate the anxiety of infinitude. Perhaps it is a working group of people with binocular vision who can remain individuals while dipping into the infinite emotionality of the group mentality. "We have to find something that is good, beautiful, and true in which we can take refuge," writes Thich Nhat Hanh in *For a Future to Be Possible*.[70] The Buddhist notion of refuge, I suspect, stands for such a containing mental space, for goodness, truth, and beauty are what it provides.

In this regard, Donald Meltzer's comments on one of his patients are thought provoking. Meltzer talks about a man who had all the requisites for happiness but was unable to be satisfied. The key to understanding this incapacity came with a dream image of a female figure riding by on a broomstick. In the dream, the patient was told to "look up quickly."[71]

This phrase meant to Meltzer's patient that the mother who might have nourished him appeared without warning and flitted by so quickly that unless he made sure to look at the right moment she was gone. His time sense narrowed to an anxious vigilance watching for these tiny openings within empty time. Outside these openings was a void, perhaps of horror, perhaps of mindlessness. The ungraspable space into which his mother seemed repeatedly to disappear, in other words, generated an anxiety in him that made him unable to receive what she was giving him, an anxiety that blotted out his capacity to assimilate and become satisfied. It was only when he could finally allow something to emerge from the past, be in the present, and disappear into the future, as it were—in other words, only by allowing the objects of his love and desire to move in their own freedom—that he could take internal satisfaction from the experience of the moment and genuinely learn from experience. In the case of this patient, the sheltering mental space of the psychoanalytic dyad allowed him to do this.

It is the conjunction of the infinite ground of the unborn state and birth into time and space that we need if the future is to be possible. Both the natural world and human experience teach this integration of what Matte Blanco would call symmetry and asymmetry. Our love of nature is particular and localized, and the pervasive living quality of the natural world manifests itself everywhere in particulars. (Blake again, seeing the infinite in a grain of sand.) At the same time, symmetries occur everywhere in natural forms. But again, as Christopher Alexander points out, they are only local, not global symmetries, and they are everywhere accompanied by slight irregularities—the quality he calls "roughness"—produced when the forces generating symmetry act within time and space.[72] A zebra's stripes may be regular in their tendency, but they must fit around the various parts of the zebra's body, which are only roughly symmetrical. The resulting combination of local symmetry and roughness, Alexander suggests, is the product of forces tending toward a larger order and integration, the equivalent in the natural world of what Bion might call mental evolution or growth.

The experience of falling in love, cited earlier as an example of symmetry, in fact manifests a similar integration of the symmetrical and the symmetrical. The notion of staying together "through thick and thin" reflects the tendency toward a larger, integrated order. We may see it as an effort to combine infiniteness and particularity. Through all thick and all thin, for as long as we live: there is the infinite. But we are two individuals with limitations who will indeed die sometime, and there will be thin times of limitation and thick times of irreplaceable specificity: there is the particularity.

Reconciled, symmetry and asymmetry manifest generativity, creativity, and unfolding. Lived time is in fact the generativity of the present contained within the penumbra of possibility. As Alexander saw, the making of something beautiful, whether in the natural or the social world—true generativity, in other words—comes about only through a process of unfolding from step to step. Working through time, it is unpredictable and uncontrollable. When we talk these days of sustainability and the seventh generation, we mean the far future. What will be sustained into that future will be not something preconceived but the continued possibility of new birth and unfolding. If extinction means an end to birth, the possible future means the continuation of generativity. As we become more capable of binocular vision and bi-logic, more capable of tolerating the frightening conjunction of symmetry and asymmetry in our experience, more capable of living in unfolding time, we become the bearers of this continuity. We become, in Octavio Paz's lovely image of regeneration and generativity, "springs of living water."

Notes

1. Thomas Berry, "The Ecozoic Era," in William Vitek and Wes Jackson, eds., *Rooted in the Land,* 195.

2. Wilfred Bion, *Clinical Seminars and Other Works,* 253.

3. See Ron Heifetz, *Leadership without Easy Answers,* esp. 22–27, 69–76.

4. Bion, "Psychiatry at a Time of Crisis," in *Cogitations,* 337–352.

5. Bion, "Psychiatry at a Time of Crisis," in *Cogitations,* 343.

6. Bion, "Psychiatry at a Time of Crisis," in *Cogitations,* 341.

7. See Bion, *Experiences in Groups,* 50.

8. See Arthur Colman, *Up from Scapegoating,* esp. 77–85.

9. Quoted in Michael Ortiz Hill, *Dreaming the End of the World,* 33. See also the discussion of Wilson's statement in chap. 5 of this volume.

10. Bion, "Psychiatry at a Time of Crisis," in *Cogitations,* 344.

11. Bion, "The 'War of Nerves,'" in Emanuel Miller and H. Crichton-Miller, eds., *The Neuroses in War,* 180.

12. Bion, "Making the Best of a Bad Job," in *Clinical Seminars and Other Works,* 322.

13. Bion, "The War of Nerves," in Miller and Crichton-Miller, eds., *The Neuroses in War,* 192.

14. Bion, *Experiences in Groups,* 48.

15. On the work group, or sophisticated group, see Bion, *Experiences in Groups,* esp. 96–99, 143. On the basic assumptions, see esp. 146–155.

16. See Bion, "On a Quotation from Freud," in *Clinical Seminars and Other Works,* 306–311.

17. Bion, "Emotional Turbulence," in *Clinical Seminars and Other Works,* 303.

18. Neil Evernden, *The Natural Alien,* 152.

19. Adolf Portmann, in Evernden, *The Natural Alien,* 121.

20. Evernden, *The Natural Alien,* 123.

21. Evernden, *The Natural Alien,* 153.

22. Evernden, *The Natural Alien,* 153.

23. Bion, "Notes on Memory and Desire," in *Cogitations,* 380–385.

24. Bion and Rickman's Northfield experiment is discussed in "Intra-Group Tensions in Therapy," included in Bion's *Experiences in Groups.* See also Malcolm Pines, ed., *Bion and Group Psychotherapy,* esp. chap. 1 (Eric Trist, "Working with Bion in the 1940s: The Group Decade") and chap. 3 (Harold Bridger, "Northfield Revisited").

25. From Bion, "The Leaderless Group Project," in *Bulletin of the Menninger Clinic,* 1946. Quoted in Trist, "Working with Bion in the 1940s," in Pines, ed., *Bion and Group Psychotherapy.*

26. Bion, "Intra-Group Tensions in Therapy," in *Experiences in Groups,* 12.

27. See Bion, *Attention and Interpretation,* chap. 12.

28. Bion, *Clinical Seminars and Other Works,* 290.

29. Bion, *Experiences in Groups,* 216.

30. See the discussions of the Northfield Experiment by Trist and Bridger in chaps. 1 and 3, respectively, in Pines, ed., *Bion and Group Psychotherapy.*

31. Bion, *All My Sins Remembered,* 58.

32. Bion, *All My Sins Remembered,* 58.

33. See Lewis Hyde, *The Gift,* chap. 3, on threshold gifts and the labor of gratitude.

34. See Hyde, *The Gift,* 48–55, for Hyde's retelling of the story of the shoemaker and the elves and his comments on it as an example of the labor of gratitude.

35. Bion, "Psychiatry at a Time of Crisis," in *Cogitations,* 351.

36. Again, this is the basic idea of Neville Symington's *Narcissism.*

37. Symington, *Narcissism,* 33.

38. Berry, "The Ecozoic Era," in Vitek and Jackson, eds., *Rooted in the Land,* 200.

39. Berry, "The Ecozoic Era," in Vitek and Jackson, eds., *Rooted in the Land,* 199.

40. Bion, *All My Sins Remembered,* 70.

41. Hill, *Dreaming the End of the World,* 75.

42. Berry, "The Ecozoic Era," in Vitek and Jackson, eds., *Rooted in the Land,* 199.

43. See C. Fred Alford, *What Evil Means to Us,* esp. chaps. 8 and 9, 117–144.

44. Laurens van der Post, *The Voice of the Thunder,* 143.

45. Harold Searles, "The Development of Mature Hope," in *Countertransference and Related Subjects,* 480.

46. Jonathan Shay, *Achilles in Vietnam,* 39.

47. See Shay, 55–67.

48. Bion, "The War of Nerves," in Miller and Crichton-Miller, eds., *The Neuroses in War,* 198.

49. Shay, *Achilles in Vietnam,* 40.

50. Christopher Lasch, quoted in Evernden, *The Natural Alien,* 151.

51. See Searles, "The Development of Mature Hope," in *Countertransference and Related Subjects,* 484.

52. See Bion, *Experiences in Groups,* 155.

53. Evernden, *The Natural Alien,* 154.

54. See Harold Boris, "On Hope: Its Nature and Psychotherapy."

55. Paul Shepard, interview in Derrick Jensen, *Listening to the Land*, 256.

56. Shay, *Achilles in Vietnam*, 180–181.

57. James Hillman, *Suicide and the Soul*, 93.

58. See Symington, *Narcissism*, chap. 8, "The Reversal of Narcissism," 81–94, esp. 92–93.

59. Hillman, *Suicide and the Soul*, 93.

60. Hillman, *Suicide and the Soul*, 93.

61. Van der Post, *The Voice of the Thunder*, 143.

62. See Margaret Mead, *Culture and Commitment*.

63. See Hillman, *The Force of Character*, 150–151.

64. See Jaques, *Requisite Organization*.

65. Ernst Bloch, "Man as Possibility," in Walter H. Capps, ed., *The Future of Hope*, 63.

66. For a general exposition of Matte Blanco's theories, see Eric Rayner, *Unconscious Logic*.

67. See Rayner, *Unconscious Logic*, 59.

68. David Harvey brilliantly expounds the complex relationship of the local and the global in his *Justice, Nature, and the Geography of Difference*.

69. Bion, *Attention and Interpretation*, 108.

70. Thich Nhat Hanh, *For a Future to Be Possible*, 7.

71. Donald Meltzer, "A Note on Introjective Processes," in *Sincerity and Other Works*, 461.

72. See Christopher Alexander's exposition of the quality of roughness in *The Nature of Order*.

Concluding Thoughts

This book has focused on the complexities of our emotional relationship to the natural world. It points to the kind of reciprocal relationships that allow for richness in growth and development, and it presents the possibility of what Bion calls *binocular vision,* in which we perceive our connection to the group and at the same time are able to reflect on that connection as separate individuals. At the same time as it provokes reflection and suggests hopeful possibilities, the book also evokes disturbing thoughts and feelings. Feeling the spur of both hope and disturbance, the reader may be wondering now where all this leaves us: What should we do? Does all this point to some course of action we should take?

Leadership and the Holding Environment

How do we achieve binocular vision in the face of emotional disturbance? To put this in terms of Symington's theory of narcissism, how do we achieve a reciprocal nurturing relationship in the face of trauma and the hopelessness and apathy it gives rise to? "If you're lucky, someone you love will pick a fight with you," said Ron Heifetz, author of *Leadership without Easy Answers,* in a workshop. He was pointing to the importance of what I have called emotional "holding" in being able to bear pain and anxiety and learn from emotional disturbance. He meant that we need to work through our conflicting loyalties and conflicting values, which will show up in our relationships with those we are close to. In other words, we need the psychological safety of a loving bond in order to dare to face our conflicts, our fears, our apathy, and our loyalties. We will need to face, at the very least, conflicts of loyalties between our human lives in our destructive society and our attachments to the natural world. Binocular vision, that is, requires a safe emotional "place" in which we can feel supported enough to notice the irrational churning away in ourselves. As Thomas Berry said, we need a collective dream—something we are emotionally invested in—in order to endure the suffering required in the transition to come.

Where are we to find this emotional holding? Originally, the loving bond was with the person who mothered us. All of us who survived infancy internalized some of the mother's capacity for "reverie"—that is, the capacity for the emotional containment of unverbalized experience. To some extent, then, we have our internal mothers, our internal containing capacities. Individual others, including the psychoanalyst or psychotherapist, may play this role for us at times as well. But the capacity to provide emotional holding is in no way limited to mothers and psychotherapists. Heifetz would say that it is a function of *leadership*: the leader's work is to provide enough safety that people can face and feel the pain of adaptive challenges without being overwhelmed by the anxieties they generate.

The leader, in other words, provides a "holding environment." But let us not be confused by the word *leader*. Leadership, as Bion and Heifetz make clear, is a function of binocular vision rather than the prerogative of a person in a position of authority. It can be exercised in many ways and by all sorts of people, in or out of positions of formal authority. Each of us has the capacity to provide some degree of a holding environment for others—and thus to provide leadership—on the basis of our own internal capacity for emotional holding.

What happens in the holding environment? Heifetz's definition of leadership is "mobilizing adaptive work"—that is, mobilizing intelligent collaboration in the service of meeting the kinds of challenges Bion was concerned with. This work involves both feeling and thinking, and it is work that leads to external action. Holding environments allow us to face the emotional issues raised in this book. Facing them, however, means not only suffering the pain of them; it also means reflecting on their meaning so that we learn from experience, and allowing this reflection to give rise to experimental action. Holding environments, in other words, are places where experiments to meet adaptive challenges are generated. Groups and meetings of all kinds, families, and the temporary intellectual community created by the internal dialogue between author and reader are instances of potential holding environments in which any participant may exercise a capacity for binocular vision at a particular point and in which new thought may arise as a result of reflection.

What should we do? We should create holding environments, and use the ones that already exist, to reflect on the issues of our relationship to the natural world and its deterioration. We should do so by exercising leadership, from whatever position we occupy, using the sources of support available to us, and with as much thoughtfulness and self-awareness, courage, and generosity as we can muster.

The Kind of Action We Call Thinking

A note on the relationship between thought and action will be helpful here. The urgency of the question "What should we do?" reflects just how disturbing it is to contemplate the emotional issues this book raises. But that urgency may lead us to a confusion about the nature of thinking. Certainly, as Bion points out, there is thinking that takes the place of action and hinders it. Thought needs to function as a preparation for action. But there is also action that substitutes for thinking, short-circuits it, if you will. We might think of violence as such a short-circuiting of thought. The wish for a clear and definitive answer to the question of what we should do reflects the urgency and anxiety we are prey to. It can be an expression of the fantasy-group mentality. In its feeling of acute helplessness, the dependency group angrily demands direction from the leader; in its eagerness to escape anxiety by rushing into action, the fight-flight group is only waiting for the enemy to be named so that it can either flee from it or rush into battle against it.

At the same time, thinking can be a form of action in itself. Where is the line between thinking, putting thoughts into words, speaking, and writing? Are they not all both thinking and acting? Formulating thoughts, putting them into words, sharing them (making them public through what Bion termed *publication*) are themselves forms of action in the larger collective. Thoughts made public reflect the binocular vision of the thinker, and in that way thought made public creates a holding environment that allows reflection by others. This book, like many others, is part of a collective thought process in which many people are participating.

Springs of Living Water: Thought, Action, and Generativity

An adaptive challenge, to use Heifetz's term, is by definition a challenge to which we do not know the answers. This means that we are of necessity experimenting when we try to address these issues. There is no certainty about the outcome. Any "action" we take will be an experiment. This is frightening in itself, of course, but it is unavoidable. It is only from experiment that we can learn. Experiments allow for reflection on outcomes and allow for further experiments. As Bion said at one point, commenting on the fact that he no longer agreed with certain views he had published earlier, at least they were good enough to get me to where I am now.

A young person once told me that because she could not decide what the best thing to do to "save the environment" was, she did not do anything. But there is no one best thing. The issue of our relation to the natural environment is so complex and multifaceted that many kinds of reflection and experimentation are useful in advancing people toward broadened concern, social responsibility, perceptual maturity, and binocular vision, to refer back to some of the concepts used in this book. And in fact, many kinds of experiments are under way, generated within many different kinds of holding environments. They may be thought of as pieces of an emergent culture.

As we saw earlier, the new cannot emerge from imitation, but it can draw on the traditions—a form of holding environment—that provide gifts to those who follow them. But this means that each person, with the help of the holding environments he or she is lucky enough to have or to help create, generates his or her own thoughts and experiments. As Heifetz points out, anyone in any position can exercise leadership, because each person in a particular role, identity, or niche in a group or society is well situated to try a specific kind of experiment on a given issue. A woman school board member is positioned to initiate, or respond, differently than a child in the classroom. Each reader of this book will, hopefully, use it toward further thought or action that makes sense in his or her own "niche." Again, it is certainly not a question of the one right or best thing to do so much as a question of allowing for binocular vision and continued reflection and experimentation wherever possible.

Let me describe this generativity a little differently. Let us suppose that one way or another we have succeeded in creating some mental space for thinking about the intense feelings connected with the natural world and its deterioration. In this space we are likely to experience stirrings of curiosity and attention. Curiosity and attention might lead us to notice configurations and connections in reality that we had been unable to see in our blind anxieties. And in this mental space we might also find ourselves more curious and attentive about our inner world, and we might find ourselves noticing creative links being made in our minds. Mental space, in short, is the generating place for creative thought, which is the basis for creative action. What should we do? We should foster the mental space that will allow for creative links to be made that can then become the basis for creative—and hence unpredictable and unprescribable—actions.

Bibliography

Abram, David. *The Spell of the Sensuous: Perception and Language in a More-Than-Human World.* New York: Pantheon, 1996.

Adorno, Theodor. *Aesthetic Theory.* Trans. Robert Hullot-Kentor. Minneapolis: University of Minnesota Press, 1997.

Aivanhov, Omraam Mikhael. Excerpt in Lucinda Vardey, ed., *God in All Worlds: An Anthology of Contemporary Spiritual Writing.* New York: Pantheon, 1995.

Alexander, Christopher. *A Foreshadowing of 21st Century Art.* New York: Oxford University Press, 1993.

———. *The Nature of Order.* New York: Oxford University Press, forthcoming.

Alford, C. Fred. *What Evil Means to Us.* Ithaca, N.Y.: Cornell University Press, 1997.

Bachelard, Gaston. *The Poetics of Space.* Trans. Maria Jolas. Boston: Beacon Press, 1969.

Bailie, Gil. *Violence Unveiled: Humanity at the Crossroads.* New York: Crossroad, 1995.

Basho, Matsu. *Narrow Road to the Interior.* Boulder: Shambhala, 1991.

Benjamin, Walter. *Illuminations.* Trans. Harry Zohn. New York: Schocken, 1969.

———. *A Berlin Childhood ca. 1900.* Unpublished translation of *Berliner Kindheit um 1900,* Shierry Weber Nicholsen, 1990.

Berman, Morris. *Wandering God: A Study in Nomadic Spirituality.* Albany, N.Y.: SUNY Press, 2000.

Berry, Thomas. *The Dream of the Earth.* San Francisco: Sierra Club Books, 1988.

———. "The Ecozoic Era." In William Vitek and Wes Jackson, eds., *Rooted in the Land: Essays on Community and Place.* New Haven: Yale University Press, 1996.

Bettelheim, Bruno. *Surviving, and Other Essays.* New York: Knopf, 1979.

Bion, Wilfred. "The War of Nerves." In Emanuel Miller and H. Crichton-Miller, eds., *The Neuroses in War.* London: Macmillan, 1940.

———. *Experiences in Groups.* London: Tavistock, 1961.

———. *Second Thoughts: Selected Papers on Psycho-Analysis.* Northvale, N.J.: Jason Aronson, 1967.

———. *Attention and Interpretation.* London: Karnac, 1984.

———. *All My Sins Remembered and The Other Side of Genius.* Oxford: Fleetwood, 1985.

———. *Clinical Seminars and Other Works.* London: Karnac, 1994.

———. *Cogitations.* London: Karnac, 1994.

Bloch, Ernst. "Man as Possibility." In Walter H. Capps, ed., *The Future of Hope.* Philadelphia: Fortress Press, 1971.

Bollas, Christopher. *Being a Character: Psychoanalysis and Self-Experience.* New York: Hill and Wang, 1992.

Boris, Harold. "On Hope: Its Nature and Psychotherapy." *International Review of Psycho-Analysis* (1976) 3, 139–150.

Brody, Hugh. *Maps and Dreams.* New York: Pantheon, 1982.

Burks, David Clarke. *Place of the Wild.* Washington, D.C.: Island Press, 1994.

Burneko, Guy. *By the Torch of Chaos and Doubt: Consciousness, Culture, Poiesis, and Religion in the Opening Global Millennium.* Cresskill, N.J.: Hampton Press, 2001.

Campbell, Joseph. *Creative Mythology.* Volume 4 of *The Masks of God.* New York: Penguin, 1968.

Chuang Tzu. *Basic Writings of Chuang Tzu.* Trans. Burton Watson. New York: Columbia University Press, 1964.

Colman, Arthur. *Up from Scapegoating: Awakening Consciousness in Groups.* Wilmette, Ill.: Chiron, 1995.

Corbett, Lionel. *The Religious Function of the Psyche.* London: Routledge, 1996.

Corbin, Henry. *Spiritual Body and Celestial Earth.* Trans. Nancy Pearson. Princeton, N.J.: Princeton University Press, 1977.

———. *The Man of Light in Iranian Sufism.* Trans. Nancy Pearson. Boulder: Shambhala, 1978.

Edinger, Edward. *Anatomy of the Psyche: Alchemical Symbolism in Psychotherapy.* LaSalle, Ill.: Open Court, 1985.

Eigen, Michael. "The Significance of the Face." In Adam Phillips, ed., *The Electrified Tightrope.* Northvale, N.J.: Jason Aronson, 1992.

Enloe, Walter, and Randy Morris. *Encounters with Hiroshima.* St. Paul, Minn.: Hamline University Press, 1998.

Evernden, Neil. *The Natural Alien: Humankind and Environment.* 2nd ed. Toronto: University of Toronto Press, 1993.

Fowles, John, and Frank Horvat. *The Tree.* Boston: Little, Brown, 1979.

Fox, Warwick. "Transpersonal Ecology and the Varieties of Identification." In Alan Drengson and Yuichi Inoue, eds., *The Deep Ecology Movement.* Berkeley: North Atlantic Books, 1995.

Gasquet, Joachim. *Joachim Gasquet's Cézanne: A Memoir with Conversations.* London: Thames and Hudson, 1991.

Griffin, Susan. *A Chorus of Stones: The Private Life of War.* New York: Doubleday, 1992.

Haines, John. "Notes from an Interrupted Journal." In David Clarke Burks, ed., *Place of the Wild,* 105–113. Washington, D.C.: Island Press, 1994.

Hanh, Thich Nhat. *For A Future To Be Possible.* Berkeley: Parallax, 1993.

Harvey, David. *Justice, Nature, and the Geography of Difference.* Malden, Mass.: Blackwell, 1996.

Heifetz, Ronald A. *Leadership Without Easy Answers.* Cambridge, Mass.: Harvard University Press, 1994.

Herman, Judith Lewis. *Trauma and Recovery.* New York: Basic Books, 1992.

Hill, Michael Ortiz. *Dreaming the End of the World: Apocalypse as a Rite of Passage.* Dallas: Spring Publications, 1994.

Hillman, James. *Suicide and the Soul.* Dallas: Spring Publications, 1973.

———. *The Dream and the Underworld.* New York: Harper & Row, 1979.

———. *The Thought of the Heart and the Soul of the World.* Dallas: Spring Publications, 1992.

———. *The Force of Character.* New York: Random House, 1999.

Holmes, Steven J. *The Young John Muir: An Environmental Biography.* Madison: University of Wisconsin Press, 1999.

Hopkins, Gerard Manley. *A Hopkins Reader.* Ed. John Pick. Garden City, N.Y.: Doubleday, 1966.

House, Freeman. *Totem Salmon: Life Lessons from Another Species.* Boston: Beacon Press, 1999.

Hyde, Lewis. *The Gift: Imagination and the Erotic Life of Property.* New York: Vintage, 1983.

Janis, Irving. "Groupthink." In David Kolb, Irwin Rubin, and James McIntyre, eds., *Organizational Psychology.* 3rd ed. Englewood Cliffs, N.J.: Prentice-Hall, 1979.

———. *Groupthink: Psychological Studies of Policy Decisions and Fiascoes.* 2nd ed. Boston: Houghton Mifflin, 1982.

Jaques, Elliot. "Death and the Mid-Life Crisis." In Elizabeth Bott Spillius, ed., *Melanie Klein Today,* vol. 1, 226–248. New York: Routledge, 1988.

———. *Requisite Organization: The CEO's Guide to Creative Structure and Leadership.* Arlington, Va.: Cason Hill, 1989.

Jensen, Derrick. *Listening to the Land: Conversations about Nature, Culture, and Eros.* San Francisco: Sierra Club Books, 1995.

Kernberg, Otto. *Love Relations.* New Haven: Yale University Press, 1995.

Lessing, Doris. *Mara and Dann.* New York: Harper, 1999.

Lifton, Robert Jay. *Death in Life: Survivors of Hiroshima.* New York: Simon and Schuster, 1967.

———. *Home from the War.* New York: Simon and Schuster, 1973.

———. *The Nazi Doctors: Medical Killing and the Psychology of Genocide.* New York: Basic Books, 1986.

Lifton, Robert Jay, and Eric Markusen. *The Genocidal Mentality: Nazi Holocaust and Nuclear Threat.* New York: Basic Books, 1990.

Lifton, Robert Jay, and Greg Mitchell. *Hiroshima in America: A Half Century of Denial.* New York: Avon, 1995.

Lindemann, Erich. *Beyond Grief: Studies in Crisis Intervention.* New York: Jason Aronson, 1979.

Livingston, John. *The Fallacy of Wildlife Conservation.* Toronto: McClelland and Stewart, 1981.

Loran, Erle. *Cézanne's Composition: Analysis of His Form, with Diagrams and Photographs of His Motifs.* Berkeley: University of California Press, [1943] 1985.

Lusseyran, Jacques. *Against the Pollution of the I: Selected Writings of Jacques Lusseyran.* New York: Parabola, 1999.

Macy, Joanna. *World as Lover, World as Self.* Berkeley: Parallax, 1991.

Mead, Margaret. *Culture and Commitment: A Study of the Generation Gap.* Garden City, N.Y.: Natural History Press, 1970.

Meltzer, Donald. *Dream-Life.* Strathclyde, Perthshire: Clunie Press, 1984.

———. *Studies in Extended Metapsychology.* Strathclyde, Perthshire: Clunie Press, 1986.

———. *Sincerity and Other Works.* London: Karnac, 1994.

Meltzer, Donald, John Bremner, Shirley Hoxter, Doreen Weddell, and Isca Wittenberg. *Explorations in Autism: A Psycho-Analytical Study.* Strathclyde, Perthshire: Clunie Press, 1975.

Meltzer, Donald, and Meg Harris Williams. *The Apprehension of Beauty.* Strathclyde, Perthshire: Clunie Press, 1988.

Merleau-Ponty, Maurice. *The Primacy of Perception.* Ed. James Edie. Evanston, Ill.: Northwestern University Press, 1964.

———. *Sense and Non-Sense.* Evanston, Ill.: Northwestern University Press, 1964.

Miller, Henry. *The Air-Conditioned Nightmare.* New York: New Directions, 1970.

Murphy, Michael, and Rhea White. *In the Zone: Transcendent Experience in Sports.* New York: Penguin/Arkana, 1995.

Nabhan, Gary Paul. "The Far Outside." In David Clarke Burks, ed., *Place of the Wild,* 19–27. Washington, D.C.: Island Press, 1994.

Nabhan, Gary Paul, and Stephen Trimble. *The Geography of Childhood: Why Children Need Wild Places.* Boston: Beacon Press, 1994.

Naess, Arne. "Identification as a Source of Deep Ecological Attitudes." In Michael Tobias, ed., *Deep Ecology,* 256–270. San Diego: Avant Books, 1985.

———. "Self-Realization: An Ecological Approach to Being in the World." In George Sessions, ed., *Deep Ecology for the 21st Century,* 225–239. Boston: Shambhala, 1995.

Noakes, J., and G. Pridham, eds. *Documents on Nazism, 1919–1945.* New York: Viking, 1975.

Ortega y Gasset, Jose. *Meditations on Hunting.* Trans. Howard Wescott. New York: Charles Scribner's Sons, 1972.

Paul, Michael Ian. *Before We Were Young: An Exploration of Primordial States of Mind.* Binghamton, N.Y.: esf Publishers, 1997.

Peacock, Doug. *Grizzly Years: In Search of the American Wilderness.* New York: Henry Holt, 1990.

Pikes, Noah. *Dark Voices: The Genesis of Roy Hart Theatre.* Woodstock, Conn.: Spring Journal Books, 1999.

Pines, Malcolm, ed. *Bion and Group Psychotherapy.* New York: Routledge, 1985.

Proctor, R. *Racial Hygiene: Medicine under the Nazis.* Cambridge, Mass.: Harvard University Press, 1988.

Raine, Kathleen. *Farewell Happy Fields.* London: Hamilton, 1973.

———. *The Land Unknown.* New York: Braziller, 1975.

Rayner, Eric. *Unconscious Logic: An Introduction to Matte Blanco's Bi-Logic and Its Uses.* London: Routledge, 1995.

Remarque, Erich Maria. *All Quiet on the Western Front.* Boston: Little, Brown, 1975.

Rhode, Eric. *On Hallucination, Intuition, and the Becoming of "O."* Binghamton, N.Y.: esf Publishers, 1998.

Rilke, Rainer Maria. *Wartime Letters of Rainer Maria Rilke.* Trans. M. D. Herter Norton. New York: Norton, 1964.

Rosen, Charles. *The Classical Style: Haydn, Mozart, Beethoven.* New York: Norton, 1972.

Ross, Lillian. *Portrait of Hemingway.* New York: Simon and Schuster, 1961.

Roszak, Theodore, Mary Gomes, and Allen Kanner, eds. *Ecopsychology.* San Francisco: Sierra Club Books, 1995.

Schachtel, Ernest. *Metamorphosis: On the Development of Affect, Perception, Attention, and Memory.* New York: Basic Books, 1959.

Searles, Harold. *The Nonhuman Environment in Normal Development and in Schizophrenia.* Madison, Conn.: International Universities Press, 1960.

———. *Countertransference and Related Subjects.* New York: International Universities Press, 1979.

Seed, John, Pat Fleming, Joanna Macy, and Arne Naess. *Thinking Like a Mountain: Towards a Council of All Beings.* Philadelphia: New Society Publishers, 1988.

Segal, Hanna. "A Psycho-analytical Approach to Aesthetics." In Melanie Klein, Paula Heimann, and R. E. Money-Kyrle, eds., *New Directions in Psychoanalysis,* 384–405. New York: Basic Books, 1955.

Sewall, Laura. *Sight and Sensibility: The Ecopsychology of Perception.* New York: Jeremy P. Tarcher/ Pentium, 1999.

Shay, Jonathan. *Achilles in Vietnam: Combat Trauma and the Undoing of Character.* New York: Athenaeum, 1994.

Shehorn, Stephen G. "The Problem of the Self in Psychology: The Psychoanalytic Self and the Self in Buddhism." Unpublished ms., 1992.

———. "The Embodied Countertransference." Unpublished ms., presented at the Black Butte conference on Psychoanalysis and the Body, 1995.

Shepard, Paul. *The Tender Carnivore and the Sacred Game.* New York: Charles Scribner's Sons, 1973.

———. *Nature and Madness.* San Francisco: Sierra Club Books, 1982.

———. "Nature and Madness." In Theodore Roszak, Mary Gomes, and Allen Kanner, eds., *Ecopsychology,* 21–40. San Francisco: Sierra Club Books, 1995.

———. *The Only World We've Got: A Paul Shepard Reader.* San Francisco: Sierra Club, 1996.

———. *The Others: How Animals Made Us Human.* Washington, D.C.: Island Press, 1996.

Snyder, Gary. *Earth House Hold.* New York: New Directions, 1969.

———. *The Practice of the Wild.* San Francisco: North Point, 1990.

———. *A Place in Space: Ethics, Aesthetics, and Watersheds.* Washington, D.C.: Counterpoint, 1995.

———. *The Gary Snyder Reader*. Washington, D.C.: Counterpoint, 1999.

Steinbeck, John. *East of Eden*. New York: Viking, 1952.

Steinbeck, John, and Edward F. Ricketts. *Sea of Cortez*. Mamaroneck, N.Y.: Viking, 1941.

Stokes, Adrian. "Form in Art." In Melanie Klein, Paula Heimann, and R. E. Money-Kyrle, eds., *New Directions in Psychoanalysis*, 406–420. New York: Basic Books, 1955.

———. *Painting and the Inner World*. London: Tavistock, 1963.

———. "The Invitation in Art." In Adrian Stokes, *Critical Writings*, vol. 3. London: Thames and Hudson, 1978.

Sullivan, Harry Stack. *The Psychiatric Interview*. New York: Norton, 1970.

Symington, Joan, and Neville Symington. *The Clinical Thinking of Wilfred Bion*. London: Routledge, 1996.

Symington, Neville. *Narcissism: A New Theory*. London: Karnac, 1993.

Tomasi di Lampedusa, Giuseppe. *The Leopard: With Two Stories and a Memoir*. Trans. Archibald Colquhoun. New York: Knopf, 1991.

Turner, Jack. "The Quality of Wildness: Preservation, Control and Freedom." In David Clarke Burks, ed., *Place of the Wild*, 175–189. Washington, D.C.: Island Press, 1994.

———. *The Abstract Wild*. Tucson: University of Arizona, 1996.

Underhill, Ruth. *The Power of Song*. Berkeley: University of California Press, 1968.

Van der Post, Laurens. *Venture to the Interior*. New York: Morrow, 1951.

———. *Yet Being Someone Other*. New York: Morrow, 1983.

———. *About Blady: A Pattern Out of Time*. New York: Morrow, 1991.

———. *The Voice of the Thunder*. New York: Morrow, 1993.

Vest, Jay Hansford C. "Sacred Geography of the Pikuni: The Badger Two Medicine Wildlands." In David Clarke Burks, ed., *Place of the Wild*, 86–93. Washington, D.C.: Island Press, 1994.

Winnicott, D. W. *The Maturational Processes and the Facilitating Environment: Studies in the Theory of Emotional Development*. London: Hogarth Press and Institute of Psycho-analysis, 1965.

———. *Playing and Reality*. London: Tavistock, 1971.

———. *Psychoanalytic Explorations*. Ed. Clare Winnicott, Ray Shepherd, and Madeleine Davis. Cambridge, Mass.: Harvard University Press, 1989.

———. *Through Paediatrics to Psycho-Analysis: Collected Papers*. New York: Brunner/Mazel, 1992.

Name Index

Abbey, Edward, 51, 73
Abram, David, 2, 4, 65, 75–77
Adorno, Theodor, 2, 31n1, 100
Aivanhov, Omraam Mikhael, 20
Alexander, Christopher, 2, 5, 95, 99, 108,
 112–113, 114, 117–118, 119, 190–191
Alford, C. Fred, 122–123, 181
Aristotle, 74

Bachelard, Gaston, 39
Bailie, Gil, 153–154
Basho, Matsu, 25
Beethoven, Ludwig van, 106–107
Benjamin, Walter, 11, 75, 104
Berman, Morris, 93n37
Berry, Thomas, 91, 94n68, 121, 122, 161,
 163, 176–177
Bettelheim, Bruno, 142–145
Bion, Wilfred, 1, 3, 5, 17, 52, 54, 74, 76, 77,
 82, 83, 84, 113, 117, 127n71, 151–152,
 155, 162, 163–164, 167–168, 170, 171–
 173, 175, 177–178, 182, 183, 186, 189,
 197
Blake, William, 66, 82, 189
Blanchot, Maurice, 84
Bloch, Ernst, 187
Bollas, Christopher, 122

Brody, Hugh, 74
Burneko, Guy, 93n35

Campbell, Joseph, 16, 96, 115
Cézanne, Paul, 4, 17, 20, 63, 66–69, 82, 84,
 107, 119
Chuang Tzu, 74
Coles, Robert, 41
Colman, Arthur, 131, 132, 165, 185
Corbin, Henri, 103–104, 118

Dante, 112
Dogen Zenji, 106, 107

Eckhart, Meister, 114–115, 121
Eigen, Michael, 97
Evernden, Neil, 9–10, 169–170, 183

Fowles, John, 22, 26, 42, 53
Fox, Warwick, 61n38
Freud, Sigmund, 110, 113, 125n40, 126n57

Gerard, Janet, 59n8
Girard, René, 153
Goethe, Johann Wolfgang von, 82
Goyen, William, 39
Griffin, Susan, 10, 12, 13, 14, 18

Subject Index